Synthesis Lectures on Mechanical Engineering

This series publishes short books in mechanical engineering (ME), the engineering branch that combines engineering, physics and mathematics principles with materials science to design, analyze, manufacture, and maintain mechanical systems. It involves the production and usage of heat and mechanical power for the design, production and operation of machines and tools. This series publishes within all areas of ME and follows the ASME technical division categories.

Mohammad H. Sadraey

Flight Stability and Control

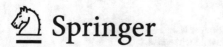 Springer

Mohammad H. Sadraey
School of Engineering, Technology,
and Aeronautics
Southern New Hampshire University
Manchester, NH, USA

ISSN 2573-3168 ISSN 2573-3176 (electronic)
Synthesis Lectures on Mechanical Engineering
ISBN 978-3-031-18767-4 ISBN 978-3-031-18765-0 (eBook)
https://doi.org/10.1007/978-3-031-18765-0

This Springer imprint is published by the registered company Springer Nature Switzerland AG
The registered company address is: Gewerbestrasse 11, 6330 Cham, Switzerland

Preface

The flight safety is a function of several parameters including stability, control, and trim. For any aircraft, to get certified, based on FAR, the controllability, stability, and trim characteristics must be shown for each flight condition within the specified flight envelope. Moreover, the aircraft must be longitudinally, directionally, and laterally stable in accordance with the provisions of airworthiness requirements. In addition, suitable stability and control feel are required in any condition normally encountered in service.

Any civil air vehicle must be stable, controllable as well as trimmable, if it is to remain in flight. Each aircraft must meet the trim, stability, and control requirements which are developed by customer and the federal government. Three concepts of trim, stability, and control are so inter-related that one must make sure all are met concurrently.

It is necessary to define suitable flight dynamic quantities to describe the flight safety requirements. This book is devoted to present the definition, fundamental parameters, basic concepts, governing equations, and requirements for trim, stability, and control. This book primarily explores stability and control of fixed-wing aircraft (not rotary-wing aircraft).

Trim is defined as the aircraft condition, when sum of all forces along each axis (x, y, and z) and sum of all moments about aircraft center of gravity about all axes are zero. *Stability* is the tendency of an aircraft to return to its initial equilibrium/trim flight condition, if disturbed. This feature allows the aircraft to remain in its present state of rest or motion despite small disturbances. *Control* is defined as a process to change the flight condition from the current trim point to a new trim point.

The flight regime of an aircraft usually includes all permissible combinations of airspeeds, altitudes, weights, centers of gravity, and configurations. This regime is primarily shaped by aerodynamics, propulsion, structure, and dynamics of aircraft. The borders of this flight regime are referred to as the flight envelope. As the flight is within the boundaries of the published flight envelope, the safety of flight is guaranteed by aircraft designer and manufacturer.

Pilots are always trained and warned through flight instruction manuals not to fly out of flight envelope, since the aircraft is either not stable, not controllable, or not structurally strong enough outside the boundaries of flight envelope. A mishap or a crash is expected, if an aircraft is flown outside the flight envelope.

The price paid for maneuverability is degrading or even the lack of stability and susceptibility to atmospheric disturbances. On the other hand, the price paid for stability is the lack of maneuverability and low controllability.

This book is written and organized into six chapters. Chapter 1 deals with aircraft equations of motion, as well as coordinate systems, aerodynamic forces and moments, and stability and control derivatives. In Chap. 2, the concept, fundamentals, and requirements of trim/balance/equilibrium are discussed. Furthermore, the governing equations of longitudinal trim, directional trim, and lateral trim are developed. Moreover, trim tab, trim wheel, trim curve, and trim plot are explained.

Chapters 3 and 4 are devoted to the concept, fundamentals, and requirements of longitudinal stability and lateral-directional stability (both static and dynamic), respectively. In Chap. 3, longitudinal state-space model, longitudinal flight modes, longitudinal characteristic equation, and neutral point are presented. Furthermore, aircraft reaction to longitudinal disturbances and aircraft components (e.g., wing, fuselage, and tail) longitudinal stability are explored.

The lateral-directional state-space model, lateral-directional characteristic equation, and lateral-directional flight modes are presented in Chap. 4. Furthermore, aircraft reaction to lateral-directional disturbances and aircraft components (e.g., wing, fuselage, tail) lateral-directional stability are explored.

Chapters 5 and 6 are devoted to the concept, fundamentals, and requirements of longitudinal control and lateral-directional control, respectively. The longitudinal transfer functions, elevator effectiveness, pitch control power, takeoff rotation control, stick force, and elevator hinge moment are also discussed in Chap. 5.

The lateral-directional transfer functions, aileron control power, turning flight, coordinated turn, rudder effectiveness, and minimum control speed are also discussed in Chap. 6. Moreover, the automatic flight control systems, autopilot, adverse yaw, and aileron reversal are briefly reviewed.

In this text, the emphasize is on the SI units or metric system (e.g., meter, kg, m/s, Newton, and Watt); since academia is moving toward this unit system. The metric unit system is taken as fundamental, this being the educational basis in all parts of the world except the USA. However, currently, all US Federal Aviation Regulations (FARs) are published in British units (e.g., slug, pound (lb), foot, nautical mile, knot, and horsepower). American (previously, British) units are still used extensively, particularly in the USA, and by industries and other federal agencies and organizations in aviation, such as FAA and NASA. Therefore, in this text, a combination of SI unit and American unit systems is utilized.

This book is intended to be used in courses such as flight dynamics, flight control systems, and flight stability and control. To help readers, in all chapters, multiple solved examples and end-of-chapter problems have been developed. Due to limited number of pages, a few topics such as "fundamentals of control systems", "flying qualities", and "dynamic modeling" are briefly covered.

I hope the reader enjoys reading this book and learns the concepts, fundamentals, and applications of stability and control in air vehicles flight. I wish to thank Springer Nature Publishers and particularly Paul Petralia, executive editor, for his support during this period. My special thanks go to Rajan Muthu—copy editor—for editing and creating an error-free text. I especially owe a large debt of gratitude to my students and the reviewers of this text. Their questions, suggestions, and criticisms have helped me to write more clearly and accurately and have influenced markedly the evolution of this book.

Manchester, USA   Mohammad H. Sadraey
August 2022

Contents

Aircraft Equations of Motion

1

1.1 Introduction

An understanding of the dynamic characteristics of an aircraft is required in assessing the flight stability and control. Flight dynamics is the science which studies the motion of an aircraft due to internally or externally generated forces or/and moments. The motion of an object can be described by means of vectors in three dimensions by using a coordinate system. To set up the equations of motion, we will use the vector analysis of classical mechanics.

This text is mainly focused on the air vehicle's stability features and control capabilities. The flight environment or atmosphere is a dynamic system which has multiple disturbances (see Fig. 1.1) to aircraft flight such as: (1) Horizontal wind, (2) Vertical wind, (3) Horizontal gust, (4) Vertical gust, (5) Wind shear, (6) Thermal column, (7) Turbulence, and (8) Sever weather. A flying aircraft must be controllable and stable, when facing these hazardous atmospheric phenomena.

The aircraft we shall address in this book is a rigid-body point mass object. To describe the complete motion of a rigid-body aircraft, one needs to consider the equations of motion with six degrees of freedom (DOF). In this chapter, the governing equations of motion for a rigid aircraft are derived. In deriving these equations, a few assumptions are made, to convert to standard forms. Moreover, the coordinate systems in which the equations are formulated are introduced.

1.2 Coordinate Axes Systems

In order to formulate the flight motion and aircraft dynamics, we need to select a coordinate system as the reference (i.e., system of coordinates or frame of reference). Coordinate

© The Author(s), under exclusive license to Springer Nature Switzerland AG 2022
M. H. Sadraey, *Flight Stability and Control*, Synthesis Lectures on Mechanical
Engineering, https://doi.org/10.1007/978-3-031-18765-0_1

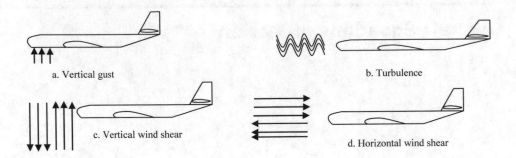

a. Vertical gust

b. Turbulence

c. Vertical wind shear

d. Horizontal wind shear

Fig. 1.1 Samples of atmospheric disturbances

system is a measurement system for locating points in space, set up within a frame of reference. We may have multiple coordinate systems within one frame of reference, each one is employed based on the flight condition and application.

Due to their desired features, Cartesian coordinate systems—that: (1) have three axes (x, y, and z), (2) are orthogonal, and (3) are right-handed—are utilized. At least two coordinate systems (Fig. 1.2) are necessary to navigate an aircraft and to study the flight dynamics: (1) Earth-fixed coordinate system (XYZ), and (2) Body-fixed coordinate system (*xyz*). The aircraft equations of motion will require the coordinate rotation relations between these two coordinate systems.

The body-fixed coordinate system is selected to be fixed to the aircraft body (i.e., fuselage reference line) and moves and rotates with it (i.e., translating and rotating). The center of this axis system is placed at the center of gravity (CG) of the aircraft. This selection which is referred to as "body-fixed axes" or simply "body axes", has an important benefit that allows us to derive the equation of motion in a desired and standard form. To form a right-handed axis system, the axis is taken with x forward, y out the right wing, and z downward, as seen from the pilot seat. Most aircraft are symmetrical with reference to the vertical plane; thus, the axes x and z lie in the plane of symmetry of the aircraft (i.e., xz plane).

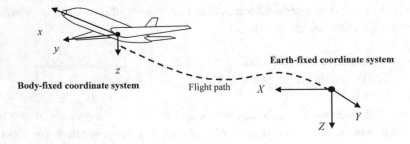

Fig. 1.2 Coordinate axes and aircraft orientation

The Earth-fixed coordinate system is employed to define the orientation of an aircraft relative to a fixed point on earth and determine navigation parameters (e.g., x, y, and z coordinates). The Earth-fixed reference frame has its z-axis pointing down, and the aircraft axes are normally aligned forward, right, and down.

The total value of x represents the flight range, and local value of z represents the aircraft altitude. For the Global Positioning System (GPS), the origin of the fixed-coordinate system is the city of Greenwich in England. However, the coordinates of a vehicle on the ground are expressed as longitude and latitude.

However, for some analyses, it is more convenient to use another axis system, (3) Wind-axes system. The lift, drag, side-force, and cross-wind forces are naturally defined in the wind-axes system. In this axis system, the x-axis is along the relative wind, or flight path. The other two axes (i.e., y and z) are following the orthogonality and right-handed rules. Note, the relative wind is often specified in north, east, and down/up components, and body-fixed components are required for aerodynamic calculations. The relative wind or aircraft airspeed is independent of the steady wind velocity.

1.3 Aerodynamic Nomenclature

There are three linear velocities (U, V, and W), and three angular velocities (P, Q, and R). Three angular velocities represent roll rate about x-axis (P), pitch rate about y-axis (Q), and yaw rate about z-axis (R). The aircraft total airspeed (V_T) has three components along x, y, and z as u, v, and w respectively. Three rotational speeds about x, y, and z are stated as p, q, and r respectively. In general, upper-case letters (e.g., U, P) are used for steady-state values, and lower-case letters (e.g., u, p) re employed for perturbed state values. Moreover, upper-case letters are used in s-domain, and lower-case letters are employed in time-domain.

Three aerodynamic forces (Fig. 1.3) along x, y, and z are stated as X, Y, and Z. However, the lift is upward, so it is along −z, and drag (D) is along the free stream, so it is along −x. Three moments about x, y, and z are referred to as rolling (L_A), pitching (M), and yawing (N) moments respectively. Moment of inertia along x, y, and z are expressed as I_{xx}, I_{yy}, and I_{zz} respectively.

Table 1.1 tabulates aerodynamic forces, moments and velocity components in body-fixed coordinate frame, as well as moments of inertia. The positive direction of three aerodynamic moments (L_A, M, and N) follows the right-hand rule (i.e., clockwise, when viewed from pilot seat, or aircraft center of gravity).

The linear velocity components u, v, and w and three group of forces will be employed is deriving three equations that govern the linear motion along x, y, and z. Similarly, three angular velocity components p, q, and r and three aerodynamic moments L_A, M, and N will be employed is deriving three equations that govern the angular motion about x, y, and z.

Fig. 1.3 Aerodynamic forces and moments

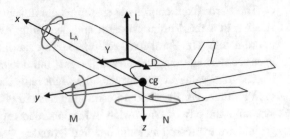

Table 1.1 Aerodynamic forces and moments and velocity components

No	Parameter	Roll axis	Pitch axis	Yaw axis
1	Axis	x	y	z
2	Angular rate	P, p	Q, q	R, r
3	Linear velocity	U, u	V, v	W, w
4	Aerodynamic force	X, D	Y	Z, L
5	Aerodynamic moment	L_A	M	N
6	Moment of inertia	I_{xx}	I_{yy}	I_{zz}
7	Product of inertia	I_{yz}	I_{xz}	I_{xy}
8	Control surface deflection	δ_A	δ_E	δ_R

The x-axis is referred to as the longitudinal axis (along the length), y-axis is referred to as the lateral axis (along the width/span), z-axis is referred to as the vertical (or normal) axis (along the height).

The conventional control surfaces sign conventions are given in Table 1.2. Since aileron has two pieces (one on the left, L, and one on the right, R), and pieces may have different deflections, the aileron deflection is determined by finding the average:

$$\delta_A = \frac{1}{2}\left(|\delta_{A_R}| + |\delta_{A_L}|\right) \tag{1.1}$$

The aerodynamic forces and moments depend on the orientation of the aircraft with respect to the relative wind. In general, two orientation aerodynamic angles are needed to specify the aerodynamic forces and moments: (1) angle of attack (α) and sideslip angle (β). The angle of attack is defined as the angle between fuselage center line and the relative wind in the xz plane. Similarly, the sideslip angle is defined as the angle between fuselage center line and the relative wind in the xy plane. This angle is positive, when the relative wind is coming from right side, as shown in Fig. 1.4.

To calculate the angle of attack, the instantaneous components of linear velocity in the directions of the x and z axes are employed:

Table 1.2 Control surfaces sign conventions

No	Control surface	Aircraft axis	Sense	Deflection	Primary effect
1	Aileron	x	+	Right-wing trailing edge down, left-wing trailing edge up	Negative rolling moment, more lift on the right-wing, less lift on the left-wing
2	Elevator	y	+	Trailing edge down	Negative pitching moment, positive tail lift
3	Rudder	z	+	Trailing edge left	Negative yawing moment

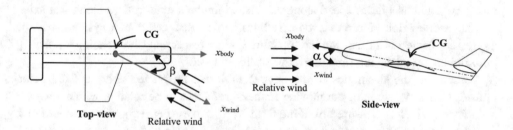

Fig. 1.4 Definition of angle of attack and sideslip angle

$$\alpha = \tan^{-1}\left(\frac{W}{U}\right) \cong \frac{W}{U} \tag{1.2}$$

where U is the airspeed along x-axis. Similarly, to calculate the sideslip angle, the instantaneous components of linear velocity in the directions of the x and y axes are employed:

$$\beta = \tan^{-1}\left(\frac{V}{U}\right) \cong \frac{V}{U} \tag{1.3}$$

The total airspeed is vector sum of three airspeed components ($V_t = \sqrt{U^2 + V^2 + W^2}$). However, for simplicity, we will normally write V throughout, instead of V_t, unless otherwise stated. Moreover, we also may use U, U_o (initial trim speed) and U_1 (steady-state speed). In Eqs. 1.2 and 1.3, α and β are in radian.

A plunging motion is when an aircraft experiences a sudden change in the flight altitude (e.g., due to a gust), while cruising. During a plunging motion, the angle of attack is changed ($\dot{\alpha}$), while the fuselage center line in the xz plane does not change. The rate of change of angle of attack ($\dot{\alpha}$) is determined by differentiating Eq. (1.2).

$$\dot{\alpha} = \frac{d\alpha}{dt} = \frac{\dot{w}}{U} \tag{1.4}$$

where, $\dot{w}$ represents the instantaneous change in the component of linear velocity in the z direction (i.e., vertical acceleration). Equations (1.2)–(1.4) are valid, if the angles of attack and sideslip are small, that is, <15°.

1.4   Vector Transformation Relationship

Now that, we have two coordinate systems (a. rotating body-fixed, b. non-rotating earth-fixed), we need to employ "vector transformation relationship". In this section, a relative-motion analysis of a rigid body using translating and rotating axes is provided. Consider an air vehicle has a linear speed along an axis, while the aircraft rotates along that axis.

The acceleration of aircraft, observed from the fixed coordinate system, may be expressed in terms of its motion measured with respect to the rotating *system of coordinates* by taking the time derivative of linear speed.

To analyze the kinematics of the aircraft subjected to rotation about a fixed point, consider the x, y, z axes of the moving frame of reference to be rotating with an angular velocity ω, which is measured from the fixed X, Y, Z axes (Fig. 1.5). The directions of i, j, and k change only on account of the rotation ω of the axis.

It states that the time derivative of any vector **A** (here, linear speed), as observed from the fixed X, Y, Z frame of reference is equal to the time rate of change of **A** as observed from the x, y, z translating-rotating frame of reference, plus $\omega \times \mathbf{A}$, the change of **A** caused by the rotation of the x, y, z frame.

$$\frac{dA}{dt}_{\,fixed-axis} = \frac{dA}{dt}_{\,moving-axis} + \omega \times A \tag{1.5}$$

For a flying aircraft, the vector is the total linear speed, V. Thus,

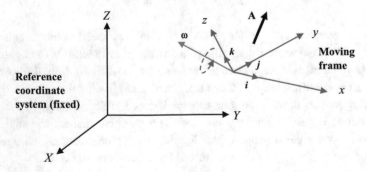

Fig. 1.5 Rotation of a moving coordinate system and a fixed frame of reference

$$\frac{dV}{dt}\bigg|_{XYZ} = \frac{dV}{dt}\bigg|_{xyz} + \omega \times V \tag{1.6}$$

In each axis, the second term represents the cross-product of the angular velocity about that axis and linear speed along that axis. The total linear velocity (V_{total}) has the following vector components:

$$V_{total} = V_x i + V_y j + V_z k = U i + V j + W k \tag{1.7}$$

The parameter ω is the total angular velocity (i.e., rotation) of the fixed coordinate and has the following vector components:

$$\omega = \omega_x i + \omega_y j + \omega_z k = P i + Q j + R k \tag{1.8}$$

The application of this equation is presented in the derivation of the aircraft equations of motion.

1.5 Aircraft Orientation and Position

In this section, the matrix algebra is employed to describe orientation and position of the aircraft and operations with coordinate systems. In order to define the orientation of an aircraft relative to the earth-fixed coordinate system, we need to define the orientation of its body axis system. The orientation of one Cartesian coordinate system (e.g., body-fixed) with respect to another (e.g., Earth-fixed) can always be described by three successive rotations by using *Euler's theorem*. The theorem states that two-component rotations about different axes passing through a point are equivalent to a single resultant rotation about an axis passing through the point.

In an aircraft (Fig. 1.3), the z-axis is pointing down and the axes are aligned forward, starboard, and down. This sequence corresponds first to a right-handed rotation around the aircraft z-axis, which is a positive "yaw". This is followed by a right-handed rotation around the aircraft y-axis, which is a positive "pitch," and a right-handed rotation around the aircraft x-axis, which is a positive "roll". Using time-varying Euler angles, the kinematic relationships for rotation in the matrix form is derived.

The order of the sequence can be chosen arbitrarily, however, the same order must be maintained ever after. In the aircraft, rotations are often performed, in a specified order, about each of the three Cartesian axes in succession. The angles of rotation (ψ, θ, ϕ) are called the *Euler angles*.

1. Clockwise rotation about the z-axis (positive ψ)
2. Clockwise rotation about the new y-axis (positive θ)
3. Clockwise rotation about the new x-axis (positive ϕ).

The rotations are stated as a yaw-pitch-roll sequence. The transformation of a vector A from Earth frame to the aircraft body frame is composed of three right-handed plane rotation matrices which each includes direction cosines:

$$A^b = \begin{bmatrix} 1 & 0 & 0 \\ 0 & \cos\phi & \sin\phi \\ 0 & -\sin\phi & \cos\phi \end{bmatrix} \begin{bmatrix} \cos\theta & 0 & -\sin\theta \\ 0 & 1 & 0 \\ \sin\theta & 0 & \cos\theta \end{bmatrix} \begin{bmatrix} \cos\psi & \sin\psi & 0 \\ -\sin\psi & \cos\psi & 0 \\ 0 & 0 & 1 \end{bmatrix} A^E \qquad (1.9)$$

where A^b and A^E denote a vector in the rotating aircraft body-axes frame and the vector in the Earth-fixed axes frame. By employing the additive property of angular velocity, and the aircraft angular velocities as the Euler angle rates, we can obtain the angular velocities in the Earth-fixed frame:

$$\begin{bmatrix} P \\ Q \\ R \end{bmatrix} = \begin{bmatrix} \dot{\phi} \\ 0 \\ 0 \end{bmatrix} + \begin{bmatrix} 1 & 0 & 0 \\ 0 & \cos\phi & \sin\phi \\ 0 & -\sin\phi & \cos\phi \end{bmatrix} \left\{ \begin{bmatrix} 0 \\ \dot{\theta} \\ 0 \end{bmatrix} + \begin{bmatrix} \cos\theta & 0 & -\sin\theta \\ 0 & 1 & 0 \\ \sin\theta & 0 & \cos\theta \end{bmatrix} \begin{bmatrix} 0 \\ 0 \\ \dot{\psi} \end{bmatrix} \right\} \qquad (1.10)$$

By multiplying out these transformations, we obtain:

$$P = \dot{\phi} - \dot{\psi} \sin\theta \qquad (1.11)$$

$$Q = \dot{\theta} \cos\phi + \dot{\psi} \cos\theta \sin\phi \qquad (1.12)$$

$$R = \dot{\psi} \cos\theta \cos\phi - \dot{\theta} \sin\phi \qquad (1.13)$$

These three coupled differential equations are called aircraft *kinematic equations*. However, to find three body-axes aircraft angular displacements of bank angle, ϕ pitch angle, θ, and yaw angle, ψ, three kinematic equations need to be solved simultaneously.

$$\dot{\phi} = P + Q \tan\theta \sin\phi + R \tan\theta \cos\phi \qquad (1.14)$$

$$\dot{\theta} = Q \cos\phi - R \sin\phi \qquad (1.15)$$

$$\dot{\psi} = \frac{1}{\cos\theta} (Q \sin\phi + R \cos\phi) \qquad (1.16)$$

Solution of these differential equations yields the components of angular velocities in body-axes relative to the earth-fixed axes (i.e., P, Q, and R). Equations (1.11)–(1.13) are body angular velocities in terms of Euler angles and Euler rates, while Eqs. (1.14)–(1.16) are Euler rates in terms body angular and velocities of Euler angles.

1.6 Aerodynamic Forces and Moments

When the air is passing around an object (here, an aircraft), the aerodynamic forces and moments are generated. The aerodynamic forces (D, Y, and L) and aerodynamic moments (L_A, M, and N) are functions of airspeed, air density, geometry, and aircraft configuration:

$$D = \frac{1}{2}\rho V^2 S C_D \tag{1.17}$$

$$Y = \frac{1}{2}\rho V^2 S C_Y \tag{1.18}$$

$$L = \frac{1}{2}\rho V^2 S C_L \tag{1.19}$$

$$L_A = \frac{1}{2}\rho V^2 S C_l b \tag{1.20}$$

$$M = \frac{1}{2}\rho V^2 S C_m C \tag{1.21}$$

$$N = \frac{1}{2}\rho V^2 S C_n b \tag{1.22}$$

where S is the wing reference (planform) area, C is the wing mean aerodynamic chord, and b is the wing span. The quantity $\frac{1}{2}\rho V^2$ has the unit of pressure and is referred to as dynamic pressure ($\bar{q}$).

$$\bar{q} = \frac{1}{2}\rho V^2 \tag{1.23}$$

The parameters, C_D, C_Y, C_L, C_l, C_m, and C_n are drag coefficient, side-force coefficient, lift coefficient, rolling moment coefficient, pitching moment coefficient, and yawing moment coefficient respectively. The aircraft configuration and flight conditions will impact the values of force and moment coefficients.

These aerodynamic forces and moments equations are skeleton of the mathematical model of aircraft aerodynamics and flight dynamics. These forces and moments will later be expressed as linearized functions of the state and control variables.

1.7 Equations of Motion

In order to analyze flight stability and control, we need to derive aircraft governing equations of motion, transfer functions, or state space representations. The equations of motion are derived by applying Newton's second law, which relates the summation of the external forces and moments to the linear and angular accelerations of the air vehicle.

We consider the aircraft as a point mass at its center of gravity (cg) as the reference point. This allows us to separate the rotational dynamics of the aircraft from the translational dynamics. This assumption will neglect any aeroelasticity effect and structural deflection, simplifies the linearization process. The moving body-axes coordinate system with a flat-earth assumption is selected.

The Newton's second law for a translational (i.e., linear) motion states that the time derivatives of linear momentum (m.V) is equal to sum of the applied forces (F).

$$\sum F = \frac{d}{dt}(mV) \tag{1.24}$$

The Newton's second law for a rotational (i.e., angular) motion states that the time derivatives of angular momentum (I.ω) is equal to sum of the applied moments (M).

$$\sum M = \frac{d}{dt}(I\omega) \tag{1.25}$$

where m, V, I, and ω are aircraft mass, total airspeed, aircraft mass moment of inertia, and angular speed respectively. Here, we assume the aircraft mass and mass moment of inertia remain constant during the period of analysis. Moreover, the mass distribution is assumed to be constant with time, which implies that the center of gravity of the aircraft remains in the same place during the interval of analysis. By using the aircraft center of gravity as a reference point, the rotational dynamics of the aircraft can be separated from the translational dynamics.

The applied moments are generated by two means: (1) aerodynamic effects, which includes the control surfaces deflections, and (2) engine thrust not acting through the center of gravity.

Since an aircraft has three axes, hence, the vehicle has six degree-of-freedom (DOF), that are three linear displacements along three axes, and three angular displacements about the three axes. The forces include aerodynamic forces (F_A) and thrust forces (T), and aircraft weight. The aerodynamic forces are X_A, Y_A, and Z_A, thrust components are X_T, Y_T, and Z_T, while L_A, M, and N are the aerodynamic moments.

Applying forces and linear velocities into Eq. (1.24) and following the kinematic relationships for rotation indicated in Eq. (1.8), we obtain:

Along x-axis:

$$X_T + X_A - W\sin(\theta) = m(\dot{U} + QW - RV) \tag{1.26}$$

Along y-axis:

$$Y_T + Y_A + W\sin(\phi)\cos(\theta) = m(\dot{V} + RU - PW) \tag{1.27}$$

Along z-axis:

$$Z_T + Z_A + W\cos(\phi)\cos(\theta) = m\big(\dot{W} + PV - QU\big) \tag{1.28}$$

Similarly, applying moments and angular velocities into Eq. (1.25), and following the kinematic relationships for rotation indicated in Eq. (1.8), we obtain:

About x-axis:

$$L_A + L_T = I_{xx}\dot{P} - I_{xz}\dot{R} - I_{xz}PQ + \big(I_{zz} - I_{yy}\big)RQ \tag{1.29}$$

About y-axis:

$$M + M_T = I_{yy}\dot{Q} + (I_{xx} - I_{zz})PR + I_{xz}\big(P^2 - R^2\big) \tag{1.30}$$

About z-axis:

$$N + N_T = I_{zz}\dot{R} - I_{xz}\dot{P} + \big(I_{yy} - I_{xx}\big)PQ + I_{xz}QR \tag{1.31}$$

In deriving Eqs. (1.26)–(1.31), we assumed a rigid air vehicle, and the effect of spinning rotors (i.e., rotating propellers and engine shaft) in the aircraft is neglected, which implies that the aircraft mass moments of inertia remain constant. In Eqs. (1.29)–(1.31), L_T, M_T, and N_T represent the rolling, pitching, and yawing moments created by the engine thrust respectively. This is due to the fact that, the thrust force may not pass through the aircraft center of gravity, or one engine is inoperative (in multi-engine aircraft).

However, to find three body-axes aircraft angular displacements of bank angle, ϕ, pitch angle, θ and yaw angle, ψ, three kinematic Eqs. (1.14), (1.15), and (1.16) need to be solved simultaneously. By integrating Eqs. (1.14), (1.15), and (1.16), the Euler angles are obtained so that the solution of the equations of motion (Eqs. 1.26–1.31) can be completed.

As Eqs. (1.29) and (1.31) indicate, the lateral motion (about x-axis) and directional motion (about z-axis) are always coupled. Any lateral motion (e.g., roll rate, P) will induce a directional motion (e.g., yaw rate, R), and any directional motion will generate a lateral motion.

By reformatting Eqs. (1.26)–(1.31), we can obtain the standard flat-earth, body-axes nonlinear coupled equations of motion which include three force- and three moment-first-order differential equations (in state-space model):

Force equations:

$$\dot{U} = RV - WQ - g\sin\theta + \frac{1}{m}[X_T + X_A] \tag{1.32}$$

$$\dot{V} = -UR + WP + g\sin\phi\cos\theta + \frac{1}{m}[Y_T + Y_A] \tag{1.33}$$

$$\dot{W} = UQ - VP + g\cos\phi\cos\theta + \frac{1}{m}[Z_T + Z_A] \tag{1.34}$$

Moment equations:

$$\dot{P} = \frac{1}{I_{xx}I_{zz} - I_{xz}^2}\left\{\left(I_{xz}(I_{xx} - I_{yy} + I_{zz})PQ\right)\right.$$
$$\left. - \left(I_{zz}(I_{zz} - I_{yy}) + I_{xz}^2\right)QR + I_{zz}L_A + I_{xz}N\right\} \tag{1.35}$$

$$\dot{Q} = \frac{1}{I_{yy}}\left[(I_{zz} - I_{xx})PR - I_{xz}(P^2 - R^2) + M\right] \tag{1.36}$$

$$\dot{R} = \frac{1}{I_{xx}I_{zz} - I_{xz}^2}\left\{\left(I_{xx}^2 + I_{xz}^2 - I_{yy}I_{xx}\right)PQ - I_{xz}(I_{xx} - I_{yy} + I_{zz})QR + I_{xz}L_A + I_{xx}N\right\}$$
$$\tag{1.37}$$

The derivation is left to the reader as a practice. The rigid-body equations of motion that were just derived is referred to as the mathematical model of aircraft dynamics.

The solution of these body-axes 6-DOF differential equations of motion will result in six motion variables of U, V, W, P, Q, and R. The primary unknowns are three linear velocities and three angular velocities. The parameters U, V, W are the linear velocity components, and P, Q, R are the corresponding angular rates. For the navigation purpose, one can integrate three linear velocities (U, V, and W) to determine the aircraft position (i.e., x, y, and z).

The aerodynamic forces along x and z in wind axis system are drag and lift (D and L). However, aerodynamic forces along body axes x and z are *axial force* (X_A) and *normal force* (Z_A). From Fig. 1.6, we can derive the following relationship:

$$Z_A = -L\cos(\alpha) - D\sin(\alpha) \tag{1.38}$$

$$X_A = L\sin(\alpha) - D\cos(\alpha) \tag{1.39}$$

In low angle of attack flight, we can have the following approximations:

$$Z_A = -L \tag{1.40}$$

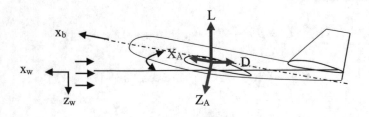

Fig. 1.6 Body axes forces and wind axes forces

$$X_A = -D \tag{1.41}$$

The thrust forces along x-axis and y-axis in body coordinate system are:

$$X_T = T \cos(\alpha + i_T) \tag{1.42}$$

$$Y_T = T \cos(\alpha) \sin(\beta) \tag{1.43}$$

where i_T is the engine installation angle in the xz plane (i.e., angle between thrust line and body x-axis). In later sections and chapter, these classifications are explored and discussed.

1.8 Simplification Techniques of Dynamic Models

The aircraft dynamic model (Eqs. 1.32–1.37) is basically a set of nonlinear coupled differential equations of motion which is based on the Newton's second law. Dealing with the nonlinear, fully coupled equations of motion can be extremely time consuming and complicated. Engineers and designers frequently resort to techniques to reduce the complexity of the equations and to make the simulation and design process easier.

When an aircraft is flying at a steady state flight conditions (i.e., when there are no linear/angular accelerations, and the control surfaces are fixed), a set of linear algebraic or differential equations may be employed for analysis. A few example cases for steady-state flights are: (1) Cruising flight ($\phi = \dot{\phi} = \dot{\theta} = \dot{\psi} = \dot{U} = \dot{V} = \dot{W} = 0$), (2) Turning flight ($\dot{\theta} = \dot{\phi} = \dot{U} = \dot{V} = \dot{W} = 0$), and (3) Pull-up ($\phi = \dot{\phi} = \dot{\psi} = \dot{U} = \dot{V} = \dot{W} = 0$). The conditions $\dot{U} = \dot{V} = \dot{W} = 0$ require the airspeed, angle of attack, and sideslip angle to be constant, and thus, the aerodynamic forces must be constant.

Two basic simplification techniques are: (1) linearization and (2) decoupling. They may be applied individually or simultaneously. Linearization is performed using either: (1) Taylor series approach, or (2) Direct method. Decoupling is performed by assuming that any rolling or yawing motion does not produce any motion in pitch, and vice versa. In other words, the longitudinal motion is independent of the lateral-directional motions.

In each case, the engineer faces different limits of application. Note that, the linearized equations of motion are valid only in the vicinity of the trim point. In other words, this is when the perturbed motion is very close to the steady state motion (trim conditions). Decoupling is also valid when the interaction between the longitudinal and lateral-directional motion can be ignored (e.g., small bank angle).

1.9   Linearization

In reality, the aircraft dynamics is nonlinear. However, dealing with a nonlinear system is not an easy task. Nonlinear systems are difficult to deal with. One way to removing or handling the nonlinearity is using a linearization technique to linearize it. The linearization converts a nonlinear model to a linear one by considering small deviation from equilibrium. Thus, the linearization is canceling or ignoring nonlinearity. Linearization is the cancellation of the nonlinearities within a system model. This Section presents the mathematical technique to linearize a nonlinear equation.

Any nonlinear equation is linearized at only an equilibrium (i.e., trim) point. The linearization is valid only at the vicinity of the equilibrium point. The trim point is a point or a motion condition that there is no acceleration (i.e., where the velocity is constant). To derive the linear equations using a Taylor series, the nonlinear aerodynamic coefficients are replaced by terms involving the stability and control derivatives.

A few flight operations where have trim points are: (1) cruising flight with a constant speed and a constant altitude, (2) climbing flight with a constant speed and a constant climb angle, and (3) turn with a constant speed and a constant radius of turn (small bank angle). Take-off and landing do not have trim conditions, since the airspeed is not constant during these two operations.

A linear system satisfies the properties of superposition and homogeneity. The general form of a linear equation is $y = mx$, where m is referred to as the slope and is a constant number. A linear term is one which is first degree in the dependent variables and their derivatives. A linear equation is an equation consisting of a sum of linear terms. The linearization about the equilibrium Point can be accomplished two methods: (1) Using Taylor series, (2) Direct substitution. Both techniques literally neglect the model's nonlinearities.

1.9.1   Taylor Series

For the case of a function with a single independent variable, by using Taylor series expansion, F(X) can be represented by the following infinite series:

$$y = f(x) = f(x_o) + \frac{df}{dx}\bigg|_{x=x_o} (x - x_o) + \frac{d^2 f}{dx^2}\bigg|_{x=x_o} \frac{(x - x_o)^2}{2!} + \cdots \quad (1.44)$$

where x_o is the trim point about which, we would like to derive the linearized equation. In linearization using Taylor series expansion, two terms are neglected: (1) higher order terms (higher than 1) and (2) constant values (e.g., $f(x_o)$). Thus, one will obtain the following linear function:

$$f(x) = \frac{df}{dx}\bigg|_{x=x_o} x \quad (1.45)$$

Basically, the linear version of a nonlinear equation (algebraic or differential) with one independent variable is equal to the slope (m) at the trim point, multiplied by the variable (f(x) = m.x). With the same token, an equation with two independent variables (say x and y) is modeled with:

$$Z = f(x, y) = \frac{df}{dx}\bigg|_{\substack{x=x_0 \\ y=y_0}} x + \frac{df}{dy}\bigg|_{\substack{x=x_0 \\ y=y_0}} y \tag{1.46}$$

where x_0 and y_0 is the values of the trim point about which, we would like to derive the linearized equation.

Example 1.1 Linearize the following function for the point (x_1, y_1, and z_1).

$$f(x, y, z) = 5z^2 y \cos(x) - 4xy$$

Solution
Taylor series:

$$f(x, y, z) = f(x_1, y_1, z_1) + \frac{\partial f}{\partial x}\bigg|_{\substack{x = x_1 \\ y = y_1 \\ z = z_1}} (x - x_1)$$

$$+ \frac{\partial f}{\partial y}\bigg|_{\substack{x = x_1 \\ y = y_1 \\ z = z_1}} (y - y_1) + \frac{\partial f}{\partial z}\bigg|_{\substack{x = x_1 \\ y = y_1 \\ z = z_1}} (z - z_1)$$

where:

$$f(x_1, y_1, z_1) = 5z_1^2 y_1 \cos(x_1) - 4x_1 y_1$$

$$\frac{\partial f}{\partial x} = -5z^2 y \sin(x) - 4y \Rightarrow \frac{\partial f}{\partial x}\bigg|_{\substack{x=x_1 \\ y=y_1 \\ z=z_1}} = -5z_1^2 y_1 \sin(x_1) - 4y_1$$

$$\frac{\partial f}{\partial y} = 5z^2 \cos(x) - 4x \Rightarrow \frac{\partial f}{\partial y}\bigg|_{\substack{x = x_1 \\ y = y_1 \\ z = z_1}} = 5z_1^2 \cos(x_1) - 4x_1$$

$$\frac{\partial f}{\partial z} = 10zy \cos(x) \Rightarrow \frac{\partial f}{\partial z}\bigg|_{\substack{x = x_1 \\ y = y_1 \\ z = z_1}} = 10z_1 y_1 \cos(x_1)$$

Substitution:

$$f(x, y, z) = 5z_1^2 y_1 \cos(x_1) - 4x_1 y_1 + [-5z_1^2 y_1 \sin(x_1) - 4y_1](x - x_1)$$
$$+ [5z_1^2 \cos(x_1) - 4x_1](y - y_1) + [10z_1 y_1 \cos(x_1)](z - z_1)$$

Expand and relocate the origin:

$$f(x, y, z) = [-5z_1^2 y_1 \sin(x_1) - 4y_1]x + [5z_1^2 \cos(x_1) - 4x_1]y + [10z_1 y_1 \cos(x_1)]z$$

This technique can be extended to the nonlinear dynamic model (equation) of an aircraft, since there are six independent variables (u, v, w, p, q, r).

1.9.2 Small-Disturbance Theory

The small-disturbance (or perturbation) theory (also referred to as direct technique) is based on the assumption, that the motion of aircraft consists of small deviations from a steady flight condition or a trim point. In applying this theory, all flight variables in a nonlinear motion equation are replaced with a reference value plus a perturbation or disturbance. This notation, after a few mathematical operations, will yield a linearized equation. This theory provides a sufficient accuracy for any flight operation will low angle of attack, and low bank angle. However, for some flight operations such as spin, steep turn, and stalling, this theorem will not have an acceptable accuracy.

In this technique, first substitute all independent variables (e.g., X) with two terms (the initial trim value; X_o, plus a small perturbation, x).

$$X = X_0 + x \tag{1.47}$$

Then, expand the non-linear equations using mathematical operations. The last step is to assume x is small, which implies retaining only linear terms (e.g., m.x). Other terms such as x^2, $\log(x)$, are removed. However, there are two exceptions. In using the small perturbation theory, for angular variables such as x, the $\sin(x)$ is replaced with x, and any $\cos(x)$ is replaced with 1.

$$\cos(x) = 1 \tag{1.48}$$

$$\sin(x) = x \tag{1.49}$$

These represent two approximations.

Example 1.2 Linearize the following first-order nonlinear differential equation (i.e., force Equation in x-direction; Eq. 5.3), using the direct substitution.

$$m(\dot{U} - VR + WQ) = X_A + X_T - mg\sin(\theta)$$

Solution
The solution is found in five steps:

1. Substitution:

$$m\left[(\dot{U}_1 + \dot{u}) - (V_1 + v)(R_1 + r) + (W_1 + w)(Q_1 + q)\right]$$
$$= -mg\sin(\Theta_1 + \theta) + (F_{AX1} + f_{Ax}) + (F_{TX1} + f_{Tx})$$

2. Expanding:

Left hand side:

$$LHS = m\left[\dot{U}_1 + \dot{U} - V_1 R_1 - V_1 r - vR_1 - vr + W_1 Q_1 + W_1 q + wQ_1 + wq\right]$$

Right hand side:

$$RHS = -mg[\sin\Theta_1\cos\theta + \sin\theta\cos\Theta_1] + F_{AX1} + f_{Ax} + F_{TX1} + f_{Tx}$$

3. Nonlinear terms (vr, wq), and constant values $(\dot{U}_1, V_1 R_1, and W_1 Q_1)$ are removed.
4. The $\sin(\theta)$ is replaced with θ, and $\cos(\theta)$ is replaced with 1 ($\sin\Theta_1\cos\theta = \sin\Theta_1$, and $\sin\theta\cos\Theta_1 = \theta\cos\Theta_1$). Since $\sin\Theta_1$ has a constant value, it is removed too.
5. Thus, the linearized equation will be:

$$m(\dot{U} - V_1 r - R_1 v + Q_1 w + W_1 q) = -mg\theta cos\Theta_1 + f_{Ax} + f_{Tx}$$

Using the small disturbance theory, all six nonlinear equations of motion are linearized as shown below:

Force equations:

$$m(\dot{U} - V_1 r - R_1 v + W_1 q + Q_1 w) = f_{Ax} + f_{Tx} - mg\cos\Theta_1\theta \qquad (1.50)$$

$$m(\dot{V} + U_1 r + R_1 u - W_1 p - P_1 w) = f_{Ay} + f_{Ty} + mg(-\sin\Phi_1\sin\Theta_1\theta$$
$$+ \cos\Phi_1\cos\Theta_1\phi) \qquad (1.51)$$

$$m(\dot{W} - U_1 q - Q_1 u + V_1 p + P_1 v) = -mg(\cos \Phi_1 \sin \Theta_1 \theta$$
$$- \sin \Phi_1 \cos \Theta_1 \phi) + f_{A_Z} + f_{T_Z} \quad (1.52)$$

where $f_{A_Y} = Y$ is the aerodynamic side force, $f_{T_Y} = T \cos(\alpha) \sin(\beta)$, $f_{T_X} = T \sin(\alpha)$, and $f_{A_Z} = -L$.

Moment equations:

$$I_{xx} \dot{P} - I_{xz} \dot{R} - I_{xz}(P_1 q + Q_1 p) + (I_{zz} - I_{yy})(R_1 q - Q_1 r) = l_A - l_T \quad (1.53)$$

$$I_{yy} \dot{Q} + (I_{xx} - I_{zz})(P_1 r + R_1 p) + I_{xz}(2P_1 p - 2R_1 r) = m_A + m_T \quad (1.54)$$

$$I_{zz} \dot{R} - I_{xz} \dot{P} + (I_{yy} - I_{xx})(P_1 q + Q_1 p) + I_{xz}(Q_1 r + R_1 q) = n_A + n_T \quad (1.55)$$

Lower case letters represent perturbed-state values of flight parameters, while upper case letters with subscript 1 represent the initial trim values of flight parameters.

1.9.3 Steady-State Equations of Motion

A steady state flight condition is defined as one for which all motion variables (mainly linear and angular velocities) remain constant with time relative to the body-axis system. When all linear and angular accelerations are removed from Eqs. (1.50)–(1.55), and all motion variables have steady state values (with a subscript 1), we obtain:
Force equations:

$$T_1 \cos(\alpha_1) - D_1 - W \sin(\theta_1) = m(Q_1 W_1 - R_1 V_1) \quad (1.56)$$

$$Y_1 + T_1 \cos(\alpha_1) \sin(\beta_1) + W \sin(\phi_1) \cos(\theta_1) = m(R_1 U_1 - P_1 W_1) \quad (1.57)$$

$$-L_1 - T_1 \sin(\alpha_1) + W \cos(\phi_1) \cos(\theta_1) = m(P_1 V_1 - Q_1 U_1) \quad (1.58)$$

Moment equations:

$$L_{A_1} + L_{T_1} = -I_{xz} P_1 Q_1 + (I_{zz} - I_{yy}) R_1 Q_1 \quad (1.59)$$

$$M_1 + M_{T_1} = (I_{xx} - I_{zz}) P_1 R_1 + I_{xz}(P_1^2 - R_1^2) \quad (1.60)$$

$$N_1 + N_{T_1} = (I_{yy} - I_{xx}) P_1 Q_1 + I_{xz} Q_1 R_1 \quad (1.61)$$

Two familiar steady state flight operations are: (1) Steady rectilinear (e.g., cruising, climbing, gliding, and descending) flight, and (2) Steady level turn. These flight cases will be explored in Chap. 2.

1.10 Decoupling

Another technique to simplify a nonlinear equation is by ignoring the coupling between some motion modes and parameters. It is widely accepted that there is subtle coupling between lateral-directional and longitudinal motions. In some flight cases (e.g., cruise), these two motions are decoupled to simplify the analysis task. In this section, first, the root causes of coupling are introduced. Then, the decoupling techniques and benefits are presented.

The motions of an aircraft are classified—due to three axes—as: (1) Longitudinal, (2) Lateral, and (3) Directional. There is a strong coupling between lateral and directional motions and state variables. Any lateral moment (e.g., aerodynamic rolling moment, L_A)—which is to generate a lateral motion—often produces a directional motion (e.g., a displacement in the y direction). On the other hand, any directional moment (e.g., yawing moment, N)—which is to generate a directional motion—often produces a lateral motion (e.g., a bank angle, ϕ). The reason lies behind the simultaneous existence of lateral moment arm and directional moment arm. In other words, any deflection of aileron (δ_A) or rudder (δ_R) will generates flight variables of β, ϕ, ψ, P, R.

- **Impact of directional motion on lateral motion**

In a conventional aircraft, the vertical tail is located on top the rear fuselage. Thus, the aerodynamic center of the vertical tail (ac_V) is above (Fig. 1.7) the center of gravity (cg) of the aircraft. Therefore, a side force that is intended to produce a yawing motion (directional motion), has a secondary effect, which is producing a rolling (lateral) motion. Thus, the directional motion cannot be decoupled from the lateral motion.

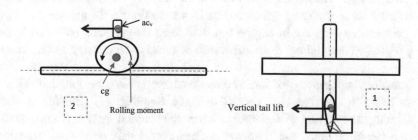

Fig. 1.7 Impact of directional motion on lateral motion

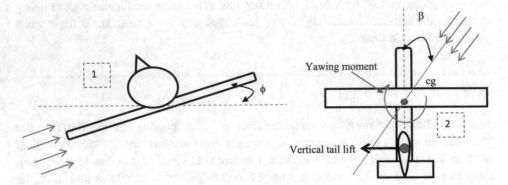

Fig. 1.8 Impact of lateral motion on directional motion

- **Impact of lateral motion on directional motion**

Assume that an aircraft is supposed to roll (Fig. 1.8). The pilot is moving the stick/wheel to deflect ailerons in order to generate a rolling motion (lateral motion). Any rolling motion produces a sideslip angle (β) too. On the other hand, a side slip angle generates a side force (Y). By the same token, this side force is producing a yawing (directional) motion. Furthermore, when both ailerons deflect with the same amount, the drag of left and right wing sections will be different. This also creates another yawing moment. Thus, the lateral motion cannot be decoupled from the directional motion.

- **Impact of lateral and directional motions on longitudinal motion**

When an aircraft has any lateral (roll), or directional (yaw) motion, it will have a number of impacts on the longitudinal motion. For instance, when an aircraft rolls (changes the bank angle; ϕ), the aircraft drag will be increased; which results in a reduction in the airspeed; V (i.e., longitudinal variable). Moreover, if the engine thrust is not used to compensate for airspeed reduction, the aircraft begins descending; or changes the altitude; h (i.e., a longitudinal variable). Another case is the influence of a change in sideslip angle (β) to an increase in the drag, which results in a reduction in the airspeed; V. Any roll or yaw requires an increase in engine thrust to keep the airspeed (compared with the cruising flight). In decoupling, these effects are assumed to be not significant, so they are neglected.

To consider the influence of lateral-directional motion on the longitudinal motion; there are basically two types of aircraft dynamic coupling: (1) coupling of the state variables, and (2) coupling of the forces. When longitudinal motion is decoupled from lateral-directional motions, the following independent longitudinal state equation and lateral-directional state equation are derived. The decoupled equations of motion may be expressed in form of state-space formulation:

$$\dot{x} = Ax + Bu$$
$$y = Cx + Du$$

(1.62)

The longitudinal state-space representation is:

$$
\begin{bmatrix} \dot{U} \\ \dot{\alpha} \\ \dot{Q} \\ \dot{\theta} \end{bmatrix} = A_{lon} \begin{bmatrix} u \\ \alpha \\ q \\ \theta \end{bmatrix} + B_{lon} \begin{bmatrix} \delta_E \\ \delta_T \end{bmatrix}
$$

(1.63)

The lateral-directional state-space representation is:

$$
\begin{bmatrix} \dot{\beta} \\ \dot{p} \\ \dot{R} \\ \dot{\phi} \end{bmatrix} = A_{lat-dir} \begin{bmatrix} \beta \\ p \\ r \\ \phi \end{bmatrix} + B_{lat-dir} \begin{bmatrix} \delta_A \\ \delta_R \end{bmatrix}
$$

(1.64)

Matrices A_{lon}, B_{lon}, $A_{lat-dir}$, and $B_{lat-dir}$ are A and B matrices for longitudinal and lateral-directional state equations respectively. These matrices will be derived in Chaps. 3–6.

These equations imply that any change in airspeed (u), angle of attack (α), and pitch angle (θ) will not induce and will not have any impact on sideslip angle (β), bank angle (ϕ), and yaw angle (ψ), and vice versa. In another word, any elevator deflection (δ_E), and throttle change (δ_T) will not generate any sideslip angle (β), bank angle (ϕ), and yaw angle (ψ). Moreover, any aileron deflection (δ_A), and rudder deflection (δ_R) will not impact airspeed (u), angle of attack (α), and pitch angle (θ).

1.11 Linearization of Forces and Moments

The change in aerodynamic forces and moments are functions of many perturbation variables including control surfaces deflection (δ_E, δ_A, and δ_R), flight variables such as linear airspeed (u, v, w), angular speeds (p, q, r, and $\dot{\alpha}$), and aircraft angles such as angle of attack (α) and sideslip angle (β). Using Taylor series expansion, one may linearize aerodynamic forces and moments in terms of the perturbation variables.

Two forces acting on the aircraft along the x body axes are the aerodynamic drag and the propulsive thrust. Two main forces acting on the aircraft along the y body axes are the aerodynamic side-force, and centrifugal force. Two main forces acting on the aircraft along the z body axes are the aerodynamic lift, and aircraft weight.

Thus, the aerodynamic forces and moments are expressed as a function of perturbation variables about the reference equilibrium condition. Using this technique, forces are represented as:

$$X = \frac{\partial X}{\partial u}u + \frac{\partial X}{\partial \alpha}\alpha + \frac{\partial X}{\partial \dot\alpha}\dot\alpha + \frac{\partial X}{\partial q}q + \frac{\partial X}{\partial \delta_E}\delta_E \tag{1.65}$$

$$Z = \frac{\partial Z}{\partial u}u + \frac{\partial Z}{\partial \alpha}\alpha + \frac{\partial Z}{\partial \dot\alpha}\dot\alpha + \frac{\partial Z}{\partial q}q + \frac{\partial Z}{\partial \delta_E}\delta_E \tag{1.66}$$

$$Y = \frac{\partial Y}{\partial \beta}\beta + \frac{\partial Y}{\partial \dot\beta}\dot\beta + \frac{\partial Y}{\partial p}p + \frac{\partial Y}{\partial r}r + \frac{\partial Y}{\partial \delta_A}\delta_A + \frac{\partial Y}{\partial \delta_R}\delta_R \tag{1.67}$$

Similarly, the aerodynamics moments are represented as:

$$L = \frac{\partial L}{\partial \beta}\beta + \frac{\partial L}{\partial p}p + \frac{\partial L}{\partial r}r + \frac{\partial L}{\partial \delta_A}\delta_A + \frac{\partial L}{\partial \delta_R}\delta_R \tag{1.68}$$

$$M = \frac{\partial M}{\partial u}u + \frac{\partial M}{\partial p}\alpha + \frac{\partial M}{\partial \dot\alpha}\dot\alpha + \frac{\partial M}{\partial q}q + \frac{\partial M}{\partial \delta_E}\delta_E \tag{1.69}$$

$$N = \frac{\partial N}{\partial \beta}\beta + \frac{\partial N}{\partial p}p + \frac{\partial N}{\partial r}r + \frac{\partial N}{\partial \delta_A}\delta_A + \frac{\partial N}{\partial \delta_R}\delta_R \tag{1.70}$$

In these equations, the nonlinear equations are linearized by terms involving the aerodynamic derivatives. All partial derivatives (e.g., $\partial M/\partial u$) have units (e.g., N/rad, Nm/rad/s) and are referred to as stability and control derivatives. In Chap. 2, we will non-dimensionalize them such that they will all have a unit of 1/rad. The numerical values of these derivatives provide clues about their relative importance. Parameters such as $\frac{\partial M}{\partial u}$, $\frac{\partial L}{\partial r}$, and $\frac{\partial N}{\partial \beta}$ are referred to as the dimensional stability derivatives, while $\frac{\partial M}{\partial \delta_E}$, $\frac{\partial L}{\partial \delta_E}$, and $\frac{\partial N}{\partial \delta_R}$ are referred to as the dimensional control derivatives. The dimensional derivatives represent either the reaction force or reaction moment applied to the aircraft, as a result of a unit disturbance in its associated motion or control variable.

The stability and control derivatives—needed for the linear equations—can be estimated relatively quickly. The linear small perturbation equations for forces and moments provide a great deal of insight into the relative importance of the various derivatives and their effect on the flight stability and control of the aircraft motion.

Since some flight variables such as airspeed and pitch rate have dimensions, they need to be multiplied/divided by some variables to make the outcome non-dimensional.

The mathematical models of linear forces and Eqs. (1.65)–(1.70) include derivations with respect different perturbations which has various units. To provide uniformity, three techniques are employed to make all variables dimensionless: (1) Speed perturbation is divided by U_1, (2) Longitudinal angular rates (i.e., pitch rate, rate of change of angle of attack) are multiplied by $\frac{\bar{c}}{2U_1}$, and (3) Lateral-directional angular rates (roll rate, yaw

Table 1.3 Non-dimensionalization of perturbation parameters for linearization

No	Perturbation type	Perturbation	Perturbation symbols	Non-dimensionalization	
				Divide/multiply perturbation by	Unitless parameter
1	Airspeed	Forward speed	u	Divide by U_1	$\frac{u}{U_1}$
2	Longitudinal angular rates	Pitch rate, rate of change of angle of attack	$q, \dot{\alpha}$	Multiply by $\frac{\bar{c}}{2U_1}$	$\frac{q\bar{c}}{2U_1}, \frac{\dot{\alpha}\bar{c}}{2U_1}$
3	Lateral-directional angular rates	Roll rate, yaw rate, rate of change of sideslip angle	$p, r, \dot{\beta}$	Multiply by $\frac{b}{2U_1}$	$\frac{pb}{2U_1}, \frac{rb}{2U_1}, \frac{\dot{\beta}b}{2U_1}$

rate, rate of change of sideslip angle) are multiplied by $\frac{b}{2U_1}$. This technique is tabulated in Table 1.3. This non-dimensionalization will change the format of Eqs. (1.65)–(1.70) as follows:

Longitudinal forces and moment:

$$X = \frac{\partial X}{\partial \left(\frac{u}{U_1}\right)}\left(\frac{u}{U_1}\right) + \frac{\partial X}{\partial \alpha}\alpha + \frac{\partial X}{\partial \left(\frac{\dot{\alpha}\bar{c}}{2U_1}\right)}\left(\frac{\dot{\alpha}\bar{c}}{2U_1}\right) + \frac{\partial X}{\partial \left(\frac{q\bar{c}}{2U_1}\right)}\left(\frac{q\bar{c}}{2U_1}\right) + \frac{\partial X}{\partial \delta_E}\delta_E \quad (1.71)$$

$$Z = \frac{\partial Z}{\partial \left(\frac{u}{U_1}\right)}\left(\frac{u}{U_1}\right) + \frac{\partial Z}{\partial \alpha}\alpha + \frac{\partial Z}{\partial \left(\frac{\dot{\alpha}\bar{c}}{2U_1}\right)}\left(\frac{\dot{\alpha}\bar{c}}{2U_1}\right) + \frac{\partial Z}{\partial \left(\frac{q\bar{c}}{2U_1}\right)}\left(\frac{q\bar{c}}{2U_1}\right) + \frac{\partial Z}{\partial \delta_E}\delta_E \quad (1.72)$$

$$M = \frac{\partial M}{\partial \left(\frac{u}{U_1}\right)}\left(\frac{u}{U_1}\right) + \frac{\partial M}{\partial \alpha}\alpha + \frac{\partial M}{\partial \left(\frac{\dot{\alpha}\bar{c}}{2U_1}\right)}\left(\frac{\dot{\alpha}\bar{c}}{2U_1}\right) + \frac{\partial M}{\partial \left(\frac{q\bar{c}}{2U_1}\right)}\left(\frac{q\bar{c}}{2U_1}\right) + \frac{\partial M}{\partial \delta_E}\delta_E \quad (1.73)$$

Lateral-directional force and moments:

$$Y = \frac{\partial Y}{\partial \beta}\beta + \frac{\partial Y}{\partial \left(\frac{\dot{\beta}\bar{c}}{2U_1}\right)}\left(\frac{\dot{\beta}\bar{c}}{2U_1}\right) + \frac{\partial Y}{\partial \left(\frac{pb}{2U_1}\right)}\left(\frac{pb}{2U_1}\right)$$

$$+ \frac{\partial Y}{\partial \left(\frac{rb}{2U_1}\right)}\left(\frac{rb}{2U_1}\right) + \frac{\partial Y}{\partial \delta_A}\delta_A + \frac{\partial Y}{\partial \delta_R}\delta_R \quad (1.74)$$

$$L = \frac{\partial L}{\partial \beta}\beta + \frac{\partial L}{\partial \left(\frac{pb}{2U_1}\right)}\left(\frac{pb}{2U_1}\right) + \frac{\partial L}{\partial \left(\frac{rb}{2U_1}\right)}\left(\frac{rb}{2U_1}\right) + \frac{\partial L}{\partial \delta_A}\delta_A + \frac{\partial L}{\partial \delta_R}\delta_R \quad (1.75)$$

$$N = \frac{\partial N}{\partial \beta}\beta + \frac{\partial N}{\partial\left(\frac{pb}{2U_1}\right)}\left(\frac{pb}{2U_1}\right) + \frac{\partial N}{\partial\left(\frac{rb}{2U_1}\right)}\left(\frac{rb}{2U_1}\right) + \frac{\partial N}{\partial \delta_A}\delta_A + \frac{\partial N}{\partial \delta_R}\delta_R \qquad (1.76)$$

Next, in three force Eqs. (1.32)–(1.34), the rate of change of velocity (i.e., linear acceleration) is multiplied aircraft mass (m), and in three moment Eqs. (1.35)–(1.37), the rate of change of angular velocity (i.e., angular acceleration) is multiplied aircraft moment of inertia (I). To keep derivatives of speeds (i.e., $\dot{U}$, $\dot{V}$, $\dot{W}$, $\dot{P}$, $\dot{Q}$, $\dot{R}$) in the left side of governing equations of motion (Eqs. 1.32–1.37), all variables on the right sides are divided by mass or moment of inertia. By employing this technique, we use new parameters for dimensional derivatives; some examples are shown below:

$$X_u = \frac{\partial X/\partial u}{m}; \quad L_q = \frac{\partial L/\partial q}{m}; \quad M_\alpha = \frac{\partial M/\partial\alpha}{I_{yy}}; \quad M_{\delta E} = \frac{\partial M/\partial\delta_E}{I_{yy}} \qquad (1.77)$$

$$Y_\beta = \frac{\partial Y/\partial\beta}{m}; \quad L_p = \frac{\partial L/\partial P}{I_{xx}}; \quad N_r = \frac{\partial N/\partial r}{I_{zz}}; \quad N_{\delta R} = \frac{\partial N/\partial\delta_R}{I_{zz}} \qquad (1.78)$$

Thus, linearized force equations will be expressed as:

$$\frac{1}{m}\Delta X = X_u u + X_\alpha \alpha + X_{\dot\alpha}\dot\alpha + X_q q + X_{\delta_E}\delta_E \qquad (1.79)$$

$$\frac{1}{m}\Delta Z = Z_u u + Z_\alpha \alpha + Z_{\dot\alpha}\dot\alpha + Z_q q + Z_{\delta_E}\delta_E \qquad (1.80)$$

$$\frac{1}{m}\Delta Y = Y_\beta\beta + Y_{\dot\beta}\dot\beta + Y_p p + Y_r r + Y_{\delta_A}\delta_A + Y_{\delta_R}\delta_R \qquad (1.81)$$

Similarly, the linearized moment equations will be expressed as:

$$\frac{1}{I_{xx}}\Delta L = L_\beta\beta + L_{\dot\beta}\dot\beta + L_p p + L_r r + L_{\delta_A}\delta_A + L_{\delta_R}\delta_R \qquad (1.82)$$

$$\frac{1}{I_{yy}}\Delta M = M_u u + M_\alpha\alpha + M_{\dot\alpha}\dot\alpha + M_q q + M_{\delta_E}\delta_E \qquad (1.83)$$

$$\frac{1}{I_{zz}}\Delta N = N_\beta\beta + N_{\dot\beta}\dot\beta + N_p p + N_r r + N_{\delta_A}\delta_A + N_{\delta_R}\delta_R \qquad (1.84)$$

These dimensional derivatives will be further expanded in Chaps. 3–6.

1.12 Stability and Control Derivatives

A group of the flight variables which are widely used in aircraft dynamic modelling are stability and control derivatives. A derivative indicates the differentiation of one parameter (e.g., x) with respect to another parameter (e.g., y). Each derivative is represented by a meaningful symbol by combining those two parameters—x as the subscript of y (here, x_y).

$$x_y = \frac{dx}{dy} \tag{1.85}$$

The stability derivatives are the rate of change (i.e., differentiation) of aerodynamic forces and moments (or their coefficients) with respect to a flight variable (e.g., angle of attack). A stability derivative represents how much change in an aerodynamic force or moment is created when there is a small change in a flight variable. These derivatives are mainly employed in the aircraft stability analysis.

In contrast, the control derivatives are simply the rate of change (i.e., differentiation) of aerodynamic forces and moments (or their coefficients) with respect to a control surface deflection (e.g., elevator). Control derivatives represent how much change in an aerodynamic force or moment is generated, when there is a small change in the deflection of a control surface. The greater a control derivative, the more powerful is the corresponding control surface. These derivatives are mainly employed in the aircraft controllability analysis.

In general, all non-dimensional derivatives begin with a capital C, have the format of C_{x_y}, and have the unit of 1/rad. Some examples of non-dimensional stability derivatives are C_{m_α}, C_{L_α}, C_{D_u}, C_{n_β}, C_{l_p}, while some examples for non-dimensional control derivatives are $C_{m_{\delta_e}}$, $C_{n_{\delta_a}}$ and $C_{l_{\delta_r}}$. The dimensional derivatives begin with a letter that represents a force (e.g., L) or moment (e.g., M), and have various units (e.g., N/rad).

A non-dimensional stability and control derivative is the partial derivative of an aerodynamic force/moment coefficient with respect to a motion variable or control variable. All non-dimensional stability and control derivatives have the same unit of 1/rad or 1/deg. Three most important non-dimensional static stability derivatives are C_{m_α} (in longitudinal static stability), C_{l_β} (in lateral static stability), and C_{n_β} (in directional static stability).

A non-dimensional control derivative is the partial derivative of an aerodynamic force/moment coefficient with respect to a control surface deflection. Three most important non-dimensional control derivatives are $C_{l_{\delta A}}$, $C_{m_{\delta E}}$, and $C_{n_{\delta R}}$. The derivative $C_{l_{\delta A}}$ is the rate of change of rolling moment coefficient with respect to a unit change in the aileron deflection. The derivative $C_{m_{\delta E}}$ is the rate of change of pitching moment coefficient with respect to a unit change in the elevator deflection. The derivative $C_{n_{\delta R}}$ is the rate of change of yawing moment coefficient with respect to a unit change in the rudder deflection.

$$C_{l_{\delta A}} = \frac{\partial C_l}{\partial \delta_A} \tag{1.86}$$

$$C_{m_{\delta E}} = \frac{\partial C_m}{\partial \delta_E} \tag{1.87}$$

$$C_{n_{\delta R}} = \frac{\partial C_n}{\partial \delta_R} \tag{1.88}$$

In aircraft handling quality evaluations, the control power requirements may be expressed and interpreted in terms the control derivatives. For instance, a rudder is designed to satisfy the directional control requirement of $C_{n_{\delta R}} < -0.4$ 1/rad for a fighter. Moreover, an elevator is designed to satisfy the longitudinal control requirement of $C_{m_{\delta E}} < -2$ 1/rad for a transport aircraft.

The aerodynamic forces and moments coefficients are modelled and linearized with respect to the stability and control derivatives. To linearize these forces and moments, there are two basics methods: (1) Linearize using non-dimensional derivatives, (2) Linearize using dimensional derivatives. Aerodynamic forces and moments coefficients are linearized with respect to non-dimensional derivatives as follows:

$$C_D = C_{D_o} + C_{D_\alpha}\alpha + C_{D_q}q\frac{C}{2U_1} + C_{D_{\dot\alpha}}\dot\alpha\frac{C}{2U_1} + C_{D_u}\frac{u}{U_1} + C_{D_{\delta_e}}\delta_e \tag{1.89}$$

$$C_y = C_{y_\beta}\beta + C_{y_p}P\frac{b}{2U_1} + C_{y_r}R\frac{b}{2U_1} + C_{y_{\delta_a}}\delta_a + C_{y_{\delta_r}}\delta_r \tag{1.90}$$

$$C_L = C_{L_o} + C_{L_\alpha}\alpha + C_{L_q}q\frac{C}{2U_1} + C_{L_{\dot\alpha}}\dot\alpha\frac{C}{2U_1} + C_{L_u}\frac{u}{U_1} + C_{L_{\delta_e}}\delta_e \tag{1.91}$$

$$C_l = C_{l_\beta}\beta + C_{l_p}P\frac{b}{2U_1} + C_{l_r}R\frac{b}{2U_1} + C_{l_{\delta_a}}\delta_a + C_{l_{\delta_r}}\delta_r \tag{1.92}$$

$$C_m = C_{m_o} + C_{m_\alpha}\alpha + C_{m_q}q\frac{C}{2U_1} + C_{m_{\dot\alpha}}\dot\alpha\frac{C}{2U_1} + C_{m_u}\frac{u}{U_1} + C_{m_{\delta_e}}\delta_e \tag{1.93}$$

$$C_n = C_{n_\beta}\beta + C_{n_p}P\frac{b}{2U_1} + C_{n_r}R\frac{b}{2U_1} + C_{n_{\delta_a}}\delta_a + C_{n_{\delta_r}}\delta_r \tag{1.94}$$

The non-dimensional stability and control derivatives will be introduced in Chaps. 2–6. It is a challenging task for a flight dynamic engineer to determine the aircraft derivatives accurately. Wind tunnel tests are a beneficial technique to calculate the stability and control derivatives.

Detailed methods for estimating the stability and control derivatives can be found in the United States Air Force Stability and Control *Datcom* [1]. The *Datcom*, short for data compendium, is a collection of methods for estimating the basic stability and control coefficients for flight regimes of subsonic, transonic, supersonic, and hypersonic speeds. Methods are presented in a systematic body build-up fashion, for example, wing alone, body alone, wing-body and wing-body-tail techniques. The methods range from techniques based on simple expressions developed from theory to correlations obtained from experimental data. Formulas for the stability derivatives as well as some limited amount of information and data on stability and control derivatives are provided in [2].

1.13 Problems

1. Linearize the following function for the point $(x_1, \text{ and } y_1)$.

$$f(x, y) = 10x^2 \cos(y) - 2xy$$

2. Linearize the following function for the point $(x_1, y_1, \text{ and } z_1)$.

$$f(x, y, z) = 3y^2x \sin(z) + 4zy$$

3. Linearize the following function for the point $(x_1, y_1, \text{ and } z_1)$.

$$f(x, y, z) = 4z^3x + 3yz \cos(x) + 14z$$

4. Linearize the following first-order nonlinear differential equation, using the direct substitution technique.

$$m(\dot{V} + RU - PW) = Y_T + Y_A + W \sin(\phi) \cos(\theta)$$

5. Linearize the following first-order nonlinear differential equation, using the direct substitution technique.

$$m(\dot{W} + PV - QU) = Z_T + Z_A + W \cos(\phi) \cos(\theta)$$

6. Linearize the following first-order nonlinear differential equation, using the direct substitution technique.

$$I_{xx}\dot{P} - I_{xz}\dot{R} - I_{xz}PQ + (I_{zz} - I_{yy})RQ = L_A + L_T$$

7. Linearize the following first-order nonlinear differential equation, using the direct substitution technique.

$$I_{yy}\dot{Q} + (I_{xx} - I_{zz})PR + I_{xz}(P^2 - R^2) = M + M_T$$

8. Linearize the following first-order nonlinear differential equation, using the direct substitution technique.

$$I_{zz}\dot{R} - I_{xz}\dot{P} + (I_{yy} - I_{xx})PQ + I_{xz}QR = N + N_T$$

References

1. Hoak D. E., Ellison D. E., et al, USAF Stability and Control DATCOM, Flight Control Division, Air Force Flight Dynamics Laboratory, Wright-Patterson AFB, Ohio, 1978
2. Etkin B., Reid L. D., Dynamics of Flight, Stability and Control, Third Edition, Wiley, 1996

Trim

2

2.1 Introduction

Trim is one of the basic requirements of a safe flight. Both stability and control are directly related and are impacted by trim. The terms trim, balance, and equilibrium are used interchangeably in the literature. Trim is defined as the aircraft condition, where sum of all forces along each axis (x, y, and z), and sum of all moments about each axis are zero. Trim is the equilibrium/balance of forces and moments in a flight operation. An aircraft is said to be trimmed, if the forces and moments acting on the aircraft are in equilibrium. When an aircraft is in trim/equilibrium, in the absence of any disturbance or input, the flight parameters stay in the same state.

Before a pilot control (makes any changes to) the aircraft, the air vehicle must be already in trim condition. The flight stability can be interpreted as "marinating a trim condition". Each flight condition with such feature is referred to as the trim point. Trim point is the condition at which all flight parameters of the aircraft have steady state values (e.g., airspeed (V), altitude (h), and angle of attack (α)). There are infinite number of trim points for an air vehicle.

An aircraft must be trimmable at all flight conditions inside the *flight envelope*. The flight envelope encompasses a number of figures; each figure is represented by allowable variations of flight parameters (e.g., aircraft weight, center of gravity, altitude, airspeed, load factor) and control parameters (e.g., elevator, aileron, and rudder deflections and engine throttle). Each flight/control parameter has a minimum value and a maximum value. These values are always published by the aircraft designer/manufacturer and must be available to the pilot. The flight safety is guaranteed, if the aircraft is flown inside the flight envelope. Figure 2.1 illustrates a typical flight envelope; it exhibits allowable variations of center of gravity along x-axis and aircraft weight. This figure is often called "weight and balance" diagram.

© The Author(s), under exclusive license to SPRINGER Nature Switzerland AG 2022
M. H. Sadraey, *Flight Stability and Control*, Synthesis Lectures on Mechanical
Engineering, https://doi.org/10.1007/978-3-031-18765-0_2

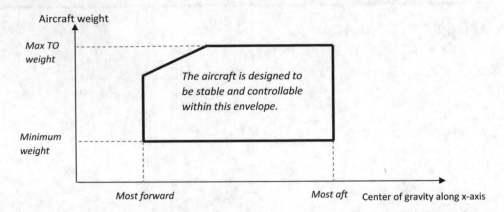

Fig. 2.1 A typical flight envelope

When an aircraft is at trim, the aircraft will not rotate about its center of gravity (cg), and aircraft will either keep moving in a desired direction (with a constant speed) or will move in a desired circular motion (with a constant angular speed). When the summations of all forces (along all three axes) and moments (about all three axes) are zero, the aircraft is in trim condition.

$$\sum F = 0 \tag{2.1}$$

$$\sum M = 0 \tag{2.2}$$

Non-aerodynamic moments (e.g., moments due to thrust) are measured with respect to the aircraft center of gravity. Any moving object (here, aircraft) has six degrees of freedom (6-DOF): (1) Three linear motions along x, y, and z axes; and (2) Three angular motions about x, y, and z axes (roll, pitch, and yaw respectively). The air vehicle trim/equilibrium must be maintained along and about three axes (x, y, and z): (1) lateral axis (x), (2) longitudinal axis (y), and (3) directional axis (z). These three trim conditions are introduced in the following three sections.

All three longitudinal, lateral, and directional trims are required throughout a flight operation. Longitudinal trim is mainly created by horizontal tail, and directional trim is primarily created by vertical tail. In a fixed-wing aircraft, longitudinal trim (in x–z plane) is almost independent of lateral (in y–z plane) and directional (in x–y plane) trim. Longitudinal trim can be mathematically modeled and analyzed without any reference to lateral and directional trim. However, lateral and directional trim are highly coupled. Any change in the lateral trim will impact the directional trim, and any change in the directional trim will impact the lateral trim. Lateral-directional trim is usually maintained simultaneously.

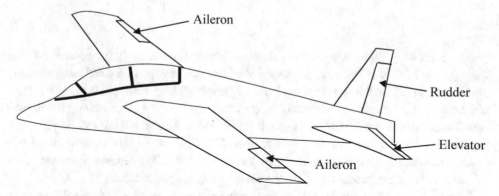

Fig. 2.2 Primary control surfaces

Reference [1] presents general trim equations and developed closed-form formulas for computing trim values for 6-DOF aircraft nonlinear dynamic models. In this technique, four inputs are elevator, aileron, rudder (Fig. 2.2), and throttle, and four trim sets are pitch angle, roll angle, angle of attack, and sideslip angle. It is interesting to note that, there are multiple solutions for each trim set requirements.

2.2   Longitudinal Trim

2.2.1   Fundamentals and Requirements

When the summation of all forces along x axis (such as drag and thrust) is zero; and the summation of all forces in z direction (such as lift and weight) is zero; and the summation of all moments—including aerodynamic pitching moment—about y axis is zero, the aircraft is said to have the longitudinal trim.

$$\sum F_x = 0 \tag{2.3}$$

$$\sum F_z = 0 \tag{2.4}$$

$$\sum M_{cg} = 0 \tag{2.5}$$

In another term, sum of all forces in the xz plane should be zero. In a low angle of attack level cruising flight, Eqs. (2.3) and (2.4) are presented in terms of lift (L), drag (D), aircraft weight (W), and engine thrust (T):

$$T \cos(\alpha) = D \tag{2.6}$$

$$W = L \tag{2.7}$$

For a cruising flight with a low angle of attack (i.e., $\alpha < 10°$), we can assume $T\cos(\alpha) = T$. This is the case for majority of cruising flights. In a fixed-wing aircraft, the horizontal tail (including elevator) is mainly responsible to maintain longitudinal trim and make the summations to be zero, by generating a necessary horizontal tail lift and contributing to the summation of moments about y axis. Horizontal tail can be placed at the rear fuselage or close to the fuselage nose. The first one is called conventional tail or aft tail, while the second one is referred to as the first tail, foreplane or canard. Two primary tools in longitudinal trim are the elevator and engine throttle.

Longitudinal trim in a conventional fixed-wing aircraft is provided mainly through the horizontal tail. To support the longitudinal trimability of the aircraft, conventional aircraft employ elevator which is part of the horizontal tail.

2.2.2 Longitudinal Trim Analysis

Consider the side view of a fixed-wing air vehicle in Fig. 2.3 that is in longitudinal trim. The figure depicts the aircraft when the aircraft center of gravity (cg) is behind the wing-fuselage aerodynamic center (ac_{wf}). There are a number of moments about y axis (i.e., CG or cg) that must be balanced by the moment of horizontal tail's lift: (1) wing-fuselage aerodynamic pitching moment, (2) the moment of lift about aircraft center of gravity, (3) Engine thrust moment, and (4) aircraft drag moment.

The aerodynamic pitching moment is the third natural outcome of any lifting surface. To follow the format of lift and drag, this moment is modelled as the product of dynamic pressure, a reference area, and a coefficient, plus the wing mean aerodynamic chord, $\overline{C}$ or simply C. This distance is conventionally chosen as the moment arm.

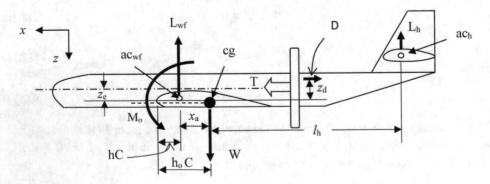

Fig. 2.3 Forces and moments in a fixed-wing aircraft in the xz plane

$$M = \frac{1}{2}\rho V^2 SCC_m \tag{2.8}$$

where C_m is the aircraft pitching moment coefficient. Three most important contributors to this aerodynamic moment are horizontal tail, wing, and fuselage. Knowledge of the pitching moment is critical to the understanding of longitudinal trim, stability and control. The aircraft pitching moment coefficient (C_m) is a function of a number of factors including pitching moment coefficient of the wing airfoil cross-section (c_m). The two-dimensional moment coefficient (c_m) is simply extracted from airfoil graphs. For instance, the value of c_m about ac for NACA airfoil 64_1-412 is -0.075.

By the application of the trim equation, and sum the moments about the center of gravity, we obtain to the following equation:

$$\sum M_{cg} = 0 \Rightarrow M_L + M_D + M_{ac} + M_T + M_{L_h} = 0 \tag{2.9}$$

In deriving this equation, a few minor moments—including the moment of tail drag, D_t—are neglected for simplicity. There are mainly five longitudinal moments about aircraft center of gravity: (1) wing-fuselage aerodynamic pitching moment (M_o or M_{ac}, simply M), (2) wing-fuselage lift moment (M_L), (3) aircraft drag moment (M_D), (4) Horizontal tail lift moment (M_{Lh}), and (5) Engine thrust's longitudinal moment (M_T).

Non-aerodynamic moments are determined by multiplying the respective force by its distance to cg. Summing these moments about the aircraft center of gravity (cg) results in:

$$Lx_a + Dz_a + M_{ac} - Tz_T - L_h l_h = 0 \tag{2.10}$$

where T and z_T are the engine thrust and distance between thrust line and cg respectively. When engine thrust line is offset from the aircraft center of gravity, the thrust will create a pitching moment (positive or negative) that must be balanced by the elevator.

Other parameters x_a, z_T, and x_{ht} are the longitudinal moment arms for wing-fuselage lift (L_{wf}), engine thrust (T), and horizontal tail lift (L_h) respectively. Note that the last two terms ($L_h \cdot l_h$ and $T \cdot z_T$) have a negative sign, since they are counterclockwise. Dividing Eq. (2.10) by $\frac{1}{2}\rho V^2 SC$, the non-dimensional three-dimensional pitching moment coefficient about the cg is obtained:

$$C_L \left(\frac{x_a}{C}\right) + C_D \left(\frac{z_d}{C}\right) + C_{m_o} - \frac{Tz_T}{\bar{q}SC} - C_{Lh} \left(\frac{S_h}{S}\right) \left(\frac{l_h}{C}\right) = 0 \tag{2.11}$$

where S_h denotes the area of the horizontal tail surface, S the wing planform area, C_{Lh} the tail lift coefficient.

Pitching moment, the torque about the aircraft center of gravity, has a profound effect on the pitch control and longitudinal stability of the air vehicle. A desired pitching moment coefficient is required to maintain trim and is obtained primarily from the tail (the last term in the equation). Any flight conditions at where the C_{mcg} is equal to zero is assumed as a longitudinal trim point. If the thrust and drag moments are negligible, Eq. (2.11) is further reduced to:

$$C_{m_o} + C_L(h - h_o) - \overline{V}_h C_{L_h} = 0 \tag{2.12}$$

The symbol $\overline{V}_h$ is the horizontal tail volume ratio:

$$\overline{V}_h = \frac{l_h S_h}{\overline{C} S} \tag{2.13}$$

where l_h is the distance between tail aerodynamic center to the aircraft center of gravity:

$$l_h = x_{ac_h} - x_{cg} \tag{2.14}$$

Typical value of $\overline{V}_h$ for a fixed-wing longitudinally stable aircraft is about 0.5–1.2. For an aircraft with a Y-tail or V-tail (e.g., Northrop Grumman RQ-4 Global Hawk UAV), tail area is the horizontal projection of the total planform area of the tail ($S_h = S_{vt} \cos(\Gamma_{vt})$); where Γ_{vt} is the V-tail dihedral angle (viewed from front view). In the Global Hawk, tail dihedral angle is set at 50°, with a 3.5 m span, an aspect ratio of 3, and each side covers an area of 4 m^2.

Two parameters h and h$_o$ are non-dimensional distances (see Fig. 2.2) between cg and the wing-fuselage ac to a reference line (often, wing leading edge at the mean aerodynamic chord location). This non-dimensional longitudinal trim equation provides a critical tool in the design of the horizontal tail.

There are three tail configurations to fulfill a change in the tail lift coefficient (C_{L_h}): (1) Fixed horizontal tail; (2) Adjustable tail; (3) All-moving tail. The fixed horizontal tail is primarily used in GA aircraft (e.g., Cessna 172), while an all-moving tail is utilized in fighter planes (e.g., McDonnell Douglas F/A-18 Hornet). An adjustable tail is mainly employed in large transport aircraft (e.g., Boeing 777 and Airbus A-350).

An adjustable tail allows the pilot to adjust its setting angle for a long time. The adjustment process usually happens before the flight; however, a pilot is allowed to adjust the tail setting angle during the flight operation. An adjustable tail employs an elevator too, but a major difference between an adjustable tail and all moving tail is in the tail rotation mechanism. The tailplane deflection for transport aircraft Boeing 777 is 4° up and 11° down.

The aerodynamic pitching moment coefficient (C_m) is a function of a number of parameters. Using Taylor series expansion, it can be expressed as a linear function of flight variables and control surface deflection as:

$$C_m = C_{m_o} + C_{m_\alpha}\alpha + C_{m_{\delta_E}}\delta_E \tag{2.15}$$

The derivatives C_{m_α} is referred to as non-dimensional stability derivative and $C_{m_{\delta_E}}$ is referred to as non-dimensional elevator control power derivative. All non-dimensional derivatives have the unit of 1/rad.

Similarly, the lift coefficient is expressed in terms of non-dimensional stability and control derivatives as:

$$C_L = C_{L_o} + C_{L_\alpha}\alpha + C_{L_{\delta_E}}\delta_E \tag{2.16}$$

where C_{L_o} is the aircraft lift coefficient at zero angle of attack. The desired horizontal tail lift coefficient is linearly modeled as:

$$C_{L_h} = C_{L_{ho}} + C_{L_{\alpha_h}}\alpha_h + C_{L_{h_{\delta E}}}\delta_E \tag{2.17}$$

The tail angle of attack is defined as:

$$\alpha_h = \alpha + i_h - \varepsilon \tag{2.18}$$

where α is the aircraft angle of attack, i_h denotes the tail incidence angle, and ε represents the downwash angle. Technique to determine elements of Eq. (2.17) is presented in Chap. 5. The parameter $C_{m_{\delta_E}}$ is referred to as the elevator control power derivative, and is determined [2] by:

$$C_{m_{\delta_E}} = -C_{L_{\alpha_h}}\overline{V}_h\eta_h\tau_e \tag{2.19}$$

where $C_{L_{\alpha_h}}$ is the horizontal tail lift curve slope and τ_e is the elevator effectiveness parameter and is expressed as the rate of change of horizontal tail angle of attack with respect to the change in elevator deflection $(d\alpha_h/d\delta_E)$. The symbol η_h is the horizontal tail efficiency (i.e., ratio of dynamic pressure at the horizontal tail (q_h) to the dynamic pressure at the wing (q)):

$$\eta_h = \frac{\overline{q}_h}{\overline{q}} = \frac{0.5\rho V_h^2}{0.5\rho V^2} = \left(\frac{V_h}{V}\right)^2 \tag{2.20}$$

Where the V is the aircraft airspeed, and the V_h is the effective airspeed at the horizontal tail region.

The local airspeed at the horizontal tail, V_h is potentially different from that at the wing-fuselage, V. Reasons for this difference may be that the tail is impacted by propeller propwash (if prop-driven engine), by exhaust gas of the engine nozzle (if jet engine), and by wing-fuselage wakes.

The elevator control power derivative is very important in takeoff rotation and landing approach.

2.2.3   Elevator Angle to Trim During Cruising Flight

When the elevator is deflected, it changes the lift, drag, and pitching moment of the aircraft. During a cruising flight (i.e., trimmed level flight), the cruise longitudinal trim requirements are reduced to the following two equations:

$$C_m = 0 = C_{m_o} + C_{m_\alpha}\alpha + C_{m_{\delta_E}}\delta_E \tag{2.21}$$

$$C_L = C_{L_c} = C_{L_o} + C_{L_\alpha}\alpha + C_{L_{\delta_E}}\delta_E \tag{2.22}$$

Two Eqs. (2.20) and (2.21) can be simultaneously solved for the elevator angle required to trim the aircraft ($\delta_{E_{trim}}$), and the aircraft cruise angle of attack (α). They are converted first into matrix form:

$$\begin{bmatrix} C_{m_\alpha} & C_{m_{\delta_E}} \\ C_{L_\alpha} & C_{L_{\delta_E}} \end{bmatrix} \begin{bmatrix} \alpha \\ \delta_E \end{bmatrix} = \begin{bmatrix} -C_{m_o} \\ C_{L_c} - C_{L_o} \end{bmatrix} \tag{2.23}$$

Then, using Cramer's rule, two unknowns are obtained. The elevator angle to trim is:

$$\delta_{E_{trim}} = \frac{\left|\begin{bmatrix} C_{m_\alpha} & -C_{m_o} \\ C_{L_\alpha} & C_{L_c} - C_{L_o} \end{bmatrix}\right|}{\left|\begin{bmatrix} C_{m_\alpha} & C_{m_{\delta_E}} \\ C_{L_\alpha} & C_{L_{\delta_E}} \end{bmatrix}\right|} = \frac{C_{m_\alpha}(C_{L_c} - C_{L_o}) + C_{L_\alpha}C_{mo}}{C_{m_\alpha}C_{L_{\delta_E}} - C_{L_\alpha}C_{m_{\delta_E}}} \tag{2.24}$$

The angle of attack ($\alpha = \alpha_c$) in a trimmed cruising flight is obtained as:

$$\alpha_c = \frac{\left|\begin{bmatrix} -C_{m_o} & C_{m_{\delta_E}} \\ C_{L_c} - C_{L_o} & C_{L_{\delta_E}} \end{bmatrix}\right|}{\left|\begin{bmatrix} C_{m_\alpha} & C_{m_{\delta_E}} \\ C_{L_\alpha} & C_{L_{\delta_E}} \end{bmatrix}\right|} = \frac{C_{m_{\delta_E}}(C_{L_o} - C_{L_c}) - C_{L_{\delta_E}}C_{m_o}}{C_{m_\alpha}C_{L_{\delta_E}} - C_{L_\alpha}C_{m_{\delta_E}}} \tag{2.25}$$

Using Eq. (2.20), one can plot variations of pitching moment coefficient (C_m) versus angle of attack (or lift coefficient) as a function of elevator deflection (see Fig. 2.4). Note that the elevator deflection does not change the slope of the pitching moment curves; it only shifts the curves, so that new trim angles of attack is achieved.

The derivative $C_{m_{\delta_E}}$ is called the elevator control power; the larger the value of $C_{m_{\delta_E}}$, the more effective the trim/control is in creating the aerodynamic pitching moment. The elevator effectiveness is proportional to the geometry of the elevator and its distance to the aircraft center of gravity. The elevator can be designed for trim requirements by considering limiting values (i.e., maximum) of elevator deflection.

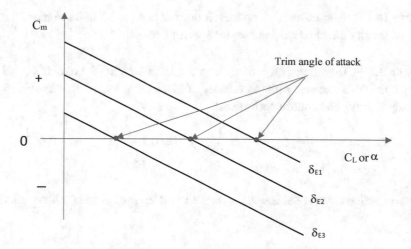

Fig. 2.4 Variations of pitching moment coefficient versus angle of attack (or lift coefficient) as a function of elevator deflection

The parameter C_{L_c} is the cruising lift coefficient and determined by equating lift and weight (Eq. 2.7), while lift is:

$$L = W = \frac{1}{2}\rho V^2 S C_L \tag{2.26}$$

Thus,

$$C_{L_c} = \frac{2W}{\rho V^2 S} \tag{2.27}$$

The largest pitching moment that must be balanced by the elevator often occurs when the aircraft is flying at the lowest speed (i.e., stall speed) and the center of gravity is at its most forward position. At this flight conditions, the maximum elevator deflection (up) is employed to provide the longitudinal trim. The variations of elevator deflection with respect to aircraft speed in order to maintain longitudinal trim in a cruising flight is often referred to as *trim curve* or *trim plot*. For this plot, the "*most aft*" aircraft center of gravity and the "*most forward*" aircraft center of gravity are considered.

The wing ac moves from about 25% MAC at subsonic speeds to about 50% MAC at supersonic speeds. This creates a challenge for supersonic aircraft designers and pilots, since aircraft needs to be longitudinally retrimmed and balance out the changes. One solution is to vary elevator deflection, and another solution is to drastically vary aircraft cg during flight. Supersonic transport aircraft *Concorde* designers chose to adjust the weight distribution of the aircraft to account for the changes in wing aerodynamic center location. Their solution was to pump fuel from front tanks to aft tanks to move the aircraft center of gravity aft, after airspeed passes speed of sound.

Example 2.1 A large transport aircraft with a takeoff weight of 190,000 lb, and a wing area of 2000 ft^2 is cruising with the velocity of 520 knot at 30,000 ft. The aircraft has the following stability and control derivatives:

$$C_{m_\alpha} = -0.9 \frac{1}{rad}, C_{L_\alpha} = 5.2 \frac{1}{rad}, C_{L_{\delta_E}} = 0.2 \frac{1}{rad}, C_{m_{\delta_E}} = -0.6 \frac{1}{rad},$$
$$C_{L_o} = 0.1, C_{mo} = -0.04$$

Determine the elevator deflection (δ_E) and aircraft angle of attack for this trimmed flight condition.

Solution
At 30,000 ft altitude, air density is 0.00089 slug/ft^3. We first need to determine the aircraft lift coefficient. In a cruising flight, lift is equal to weight:

$$C_L = C_{L_c} = \frac{2W}{\rho V^2 S} = \frac{2 \times 190,000}{0.00089 \times (520 \times 1.688)^2 \times 2000} = 0.277 \qquad (2.27)$$

where 1.688 is to convert knot to ft/s. The elevator deflection:

$$\delta_{E_{trim}} = \frac{C_{m_\alpha}(C_{L_c} - C_{L_o}) + C_{L_\alpha} C_{mo}}{C_{m_\alpha} C_{L_{\delta_E}} - C_{L_\alpha} C_{m_{\delta_E}}} \qquad (2.24)$$

$$\delta_{E_{trim}} = \frac{-0.9 \times (0.277 - 0.1) + 5.2 \times (-0.04)}{(-0.9) \times 0.2 - 5.2 \times (-0.6)} = -0.125 \, rad = -7.16 \, deg$$

The aircraft angle of attack:

$$\alpha_c = \frac{C_{m_{\delta_E}}(C_{L_o} - C_{L_c}) - C_{L_{\delta_E}} C_{mo}}{C_{m_\alpha} C_{L_{\delta_E}} - C_{L_\alpha} C_{m_{\delta_E}}} \qquad (2.25)$$

$$\alpha_c = \frac{(-0.6) \times (0.277 - 0.1) - 0.2 \times (-0.04)}{(-0.9) \times 0.2 - 5.2 \times (-0.6)} = 0.054 \, rad = 3.08 \, deg$$

2.2.4   Elevator Angle to Trim During Climbing Flight

Figure 2.5 illustrates the forces and moments on an aircraft in a steady-state constant-speed climbing flight. During a climbing flight (i.e., trimmed flight, while climbing), the longitudinal trim requirements are governed by Eqs. (2.3)–(2.5). The summation of all

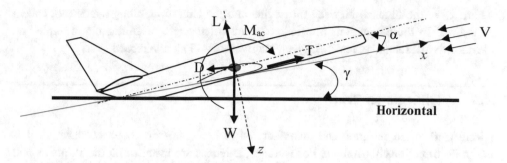

Fig. 2.5 Forces and moments diagram for a climbing flight

forces (along x and z) and moments about y axis (cg) must be equal to zero. By inserting forces and moments, we obtain:

$$\sum F_x = 0 \Rightarrow T\cos(i_T + \alpha) - D - W\sin(\gamma) = 0 \tag{2.28}$$

$$\sum F_z = 0 \Rightarrow L + T\sin(i_T + \alpha) - W\cos(\gamma) = 0 \tag{2.29}$$

$$\sum M_{cg} = 0 \Rightarrow \frac{1}{2}\rho V^2 SCC_m + Tz_T = 0 \tag{2.30}$$

The aircraft engine often has a setting angle, i_T (about 2–5°). If we ignore the engine setting angle in our calculations, the thrust is acting parallel to the fuselage center line. In general, this is not quite true, but in conventional aircraft, the effects of an inclination of the thrust vector are small enough to be neglected. For simplicity, we assume the thrust line is in the direction of flight (ignoring the effect of angle of attack on engine thrust). Furthermore, we assume aerodynamic forces are passing through the aircraft center of gravity.

$$\frac{1}{2}\rho V^2 S\overline{C}\left(C_{m_o} + C_{m_\alpha}\alpha + C_{m_{\delta_E}}\delta_E\right) + Tz_T = 0 \tag{2.31}$$

$$\frac{1}{2}\rho V^2 S\left(C_{L_o} + C_{L_\alpha}\alpha + C_{L_{\delta_E}}\delta_E\right) - W\cos(\gamma) = 0 \tag{2.32}$$

$$\frac{1}{2}\rho V^2 S\left(C_{D_o} + C_{D_\alpha}\alpha + C_{D_{\delta_E}}\delta_E\right) + W\sin(\gamma) = T\cos(\alpha) \tag{2.33}$$

where z_T is the distance between thrust line and the aircraft cg along z-axis (not shown in figure). In these equations, the unknowns are often angle of attack, climb angle and elevator deflection (α, γ, and δ_E), given engine thrust (T) and airspeed (V).

2.3   Directional Trim

Aircraft are often designed and manufactured to be symmetric about xz plane, so it is naturally directionally trimmed. For instance, antennas are installed on the xz plane, and pitot tubes are either installed on the xz plane or on both sides of the aircraft. Moreover, in a twin-engine aircraft, engines are installed on two sides of the fuselage (see Fig. 2.6) with the same distance from fuselage centerline. Thus, the yawing moments of thrust forces of left and right engine are nullified. If one engine becomes inoperative, the rudder will be deflected to balance/trim the asymmetric thrust moment. Another example for the prop-driven twin-engine aircraft is that the rotation of propellers on each side are different (i.e., one clockwise, and another one is counter-clockwise).

However, in some unusual flight conditions, the aircraft may become asymmetric about xz plane. A few instances of asymmetricity about xz plane are: (1) When a fighter aircraft launches one missile from one side of the aircraft, the missile on the other side will create an extra drag which causes a yawing moment. (2) If the fuel valve of a fuel tank in one side of an aircraft (say, left tank) has malfunction, the aircraft only will consume fuel from another side (say, right tank). This causes one side to be heavier than the other side, and this asymmetricity about xz plane requires aileron deflection for lateral trim. This incidence consequently generates an adverse yaw that requires a directional trim. Due to this process, both aileron and rudder (δ_A and δ_R) are simultaneously used in the directional trim.

When the summation of all forces along y axis (e.g., centrifugal force and side force) is zero; and the summation of all moments including aerodynamic yawing moment about z axis (i.e., N) is zero, the aircraft is said to have the directional trim.

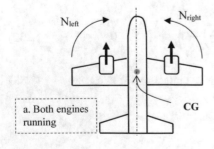 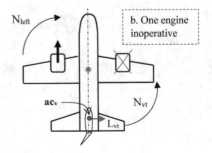

Fig. 2.6 Directional trim

$$\sum F_y = 0 \tag{2.34}$$

$$\sum N_{cg} = 0 \tag{2.35}$$

In a fixed-wing aircraft, the vertical tail (including rudder) is mainly responsible to maintain directional trim and to make the summations to be zero, by generating a necessary vertical tail lift (to the left or right).

The most important contributor to the aerodynamic yawing moment is the vertical tail (through rudder). A rudder deflection initially generates a vertical tail lift, which consequently creates a yawing moment. Furthermore, it generates a sideslip angle which results in a side force. The third unintended production of the rudder deflection is a rolling moment. This is due to the fact that the vertical tail is frequently placed above the aircraft cg.

In a symmetric aircraft with a zero sideslip angle, and a zero aileron deflection, the aerodynamic yawing moment is determined by multiplying the vertical tail lift by the vertical tail arm:

$$N_A = L_V \cdot l_v \tag{2.36}$$

where l_v is the vertical tail arm and is the distance; along x axis; between aircraft cg and vertical tail aerodynamic center (ac_v). The vertical tail aerodynamic center is usually located at the quarter chord of the vertical tail mean aerodynamic chord.

In a level cruising flight, there is normally no force generated along the y direction. However, in a cross-wind landing (see Fig. 2.7), the aircraft must maintain the direction (i.e., directional trim) along runway, while the wind creates a sideslip angle. In such condition, both aileron and rudder are deflected to simultaneously maintain lateral and directional trim. There are two specific techniques for cross-wind landing: (1) Crabbing, and (2) Side slipping. With both techniques, pilot can maintain the alignment with the runway centerline, each has pros and cons. The necessary crab angle is determined by the strength of the crosswind component.

To keep the flight direction or heading angle (ψ), the aircraft must use rudder, which often creates a bank angle. In this condition, there are often two active forces in the horizontal (xy) plane: (1) Aerodynamic side force (F_y), and (2) Horizontal component of the lift ($L \sin (\phi)$). When an aircraft banks with a bank angle (ϕ), the lift has two components: (1) The horizontal component ($L \sin (\phi)$), and (2) Vertical component to balance the aircraft weight:

$$\sum F_z = 0 \Rightarrow L \cos \phi = W \tag{2.37}$$

The aerodynamic side force is a function of air density (i.e., altitude), wing area, airspeed, and side force coefficient (C_y).

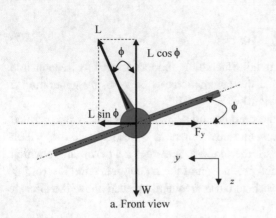

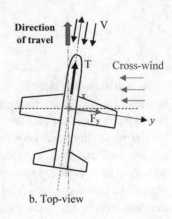

b. Top-view

a. Front view

Fig. 2.7 Directional trim in cross-wind landing

$$F_y = \frac{1}{2}\rho V^2 S C_y \tag{2.38}$$

The side force is positive when it is along y axis (starboard). The side force coefficient (C_y) is a function of aircraft configuration (C_{yo}); as well as sideslip angle (β), rudder deflection (δ_R), and aileron deflection (δ_A) and yaw rate (r). Using Taylor series expansion, it can be expressed as a linear function of flight variables and control surfaces. By employing short-hand expression and non-dimensionalization technique, this equation is presented as:

$$C_y = C_{yo} + C_{y\beta}\beta + C_{y\delta_A}\delta_A + C_{y\delta_R}\delta_R \tag{2.39}$$

where $C_{y\beta}$ is an aircraft stability derivative, and $C_{y\delta_A}$ and $C_{y\delta_R}$ are two aircraft control derivatives. When forces along y-axis are substituted into Eq. (2.34), the following result is obtained:

$$L\sin\phi + F_y = 0 \tag{2.40}$$

The aerodynamic yawing moment is:

$$N = \frac{1}{2}\rho V^2 S C_n b \tag{1.22}$$

where C_n is the aircraft yawing moment coefficient, and b denotes the wing span. The aircraft yawing moment coefficient (C_n) is a function of a number of flight and control surface parameters. Using Taylor series expansion, it can be expressed as a linear function of flight variables and control surfaces. By employing short-hand expression and non-dimensionalization technique, this equation is presented as:

$$C_n = C_{n_o} + C_{n_\beta}\beta + C_{n_{\delta A}}\delta_A + C_{n_{\delta R}}\delta_R \tag{2.41}$$

where C_{n_β} is another non-dimensional stability derivative, and $C_{n_{\delta A}}$ and $C_{n_{\delta R}}$ are two non-dimensional control derivatives. You may assume that the aircraft is symmetric about xz plane (i.e., $C_{n_o} = 0$).

The parameter $C_{n_{\delta R}}$ is referred to as the rudder control power derivative, and is determined [2] by:

$$C_{n_{\delta R}} = -C_{L_{\alpha V}}\overline{V}_V \eta_V \tau_r \tag{2.42}$$

where $C_{L_{\alpha V}}$ is the vertical tail lift curve slope and η_V is the vertical tail efficiency (i.e., ratio of dynamic pressure at the vertical tail to the dynamic pressure at the wing; q_v/q). The parameter τ_r is the rudder effectiveness parameter and is expressed as the rate of change of vertical tail angle of attack with respect to the change in rudder deflection ($d\alpha_v/d\delta_R$). Finally, $\overline{V}_V$ is the vertical tail volume ratio:

$$\overline{V}_V = \frac{l_v S_v}{bS} \tag{2.43}$$

where b is wingspan, and S_v is the vertical tail area. Typical value of $\overline{V}_v$ for a fixed-wing directionally stable aircraft is about 0.02–0.09. The rudder control power derivative is very important in helping to overcome asymmetric thrust situations, and in coordinating turns.

Another important case of directional trim for a multi-engine aircraft is, when the aircraft is in one-engine-inoperative (OEI) condition. An aircraft must be directionally trimmable for the appropriate flight phases (Fig. 2.7), when any single failure or malfunction of the propulsive system, including inlet or exhaust, causes loss of thrust on one or more engines or propellers, considering also the effect of the failure or malfunction on all subsystems powered or driven by the failed propulsive system.

Given the aircraft geometry and engine thrust, one can calculate the rudder deflection to keep aircraft directionally trimmed. The maximum rudder deflection is required when aircraft has the lowest airspeed, while the center of gravity is at its most aft position, and the operative engine is generating its maximum thrust. The velocity above which an aircraft is not directionally trimmable is referred to as the minimum control speed. This speed will be further explored in Chap. 6.

It must be possible to take off and land with normal pilot skill and technique in 90° crosswinds, from either side, of velocities up to a specified value. According to [3], the class I aircraft are required to handle a cross wind landing with a maximum wind speed of 20 knot. For class II, III, and IV, the maximum cross-wind speed requirement is 30 knot.

The rudder must be powerful enough to maintain directional trim in a crosswind take-off/landing operation. For all airplanes except land-based airplanes equipped with crosswind landing gear, or otherwise constructed to land in a large crabbed attitude, yaw- and roll-control power shall be adequate to develop at least 10° of sideslip in the power approach with yaw control pedal forces not exceeding some specified values as given in [3].

Example 2.2 A large transport aircraft with a maximum take-off weight of 140,000 lb is equipped with two turbofan engines each generating 26,000 lb of thrust. Right after takeoff—when airspeed is 120 knot—right engine becomes inoperative. The lateral distance between two engines is 40 ft. Other characteristics of the aircraft are as follows:

$$C_{L_{\alpha V}} = 4.3 \frac{1}{\text{rad}}; \, S = 1200\,\text{ft}^2; \, b = 100\,\text{ft}; \, S_V = 250\,\text{ft}^2; \, l_v = 65\,\text{ft}, \, \tau_r = 0.5, \, \eta_v = 0.98$$

Determine how much rudder deflection is needed (and to which direction) to directionally trim the aircraft.

Solution
The vertical tail volume coefficient is:

$$\overline{V}_V = \frac{l_V S_V}{bS} = \frac{65 \times 250}{100 \times 1200} = 0.135 \tag{2.43}$$

$$C_{n_{\delta_R}} = -C_{L_{\alpha V}} \overline{V}_V \eta_V \tau_r = -4.3 \times 0.135 \times 0.98 \times 0.5 = -0.282 \frac{1}{\text{rad}} \tag{2.42}$$

At directional trim, sum of the directional moments is zero.

$$\sum N_{cg} = 0 \tag{2.35}$$

There is a yawing moment created by asymmetric thrust:

$$N_T = T_R y_T = 26{,}000 \times \frac{40}{2}$$

where T_R stands for thrust of the right engine. Another moment is the aircraft aerodynamic moment:

$$N = \frac{1}{2}\rho V^2 SbC_n \tag{1.22}$$

where the yawing moment coefficient is:

$$C_n = C_{n_o} + C_{n_\beta}\beta + C_{n_{\delta_A}}\delta_A + C_{n_{\delta_R}}\delta_R \tag{2.41}$$

However, there is no sideslip angle, so:

$$C_n = C_{n_{\delta_R}}\delta_R$$

By inserting this C_n into Eq. (1.22), and then, two moments back into Eq. (2.35), the rudder deflection to balance the asymmetric thrust at sea level is:

$$\delta_R = \frac{T_L y_T}{-\frac{1}{2}\rho V^2 SbC_{n_{\delta_R}}} = \frac{26{,}000 \times \frac{40}{2}}{-\frac{1}{2} \times 1.225 \times (130 \times 1.688)^2 \times 1200 \times 100 \times (-0.282)}$$

$$\delta_R = 0.315 \, \text{rad} = 18.03 \, \text{deg}$$

The rudder must be deflected to the left to balance the yawing moment of the left engine.

2.4 Lateral Trim

Aircraft are often designed and manufactured to be symmetric about xz plane, so it is naturally laterally trimmed. For instance, both left and right wing (and horizontal tail) sections are symmetric.

However, in some unusual flight conditions, the aircraft may become asymmetric about xz plane. One instance of asymmetricity about xz plane is, if the fuel valve of a fuel tank in one side of an aircraft (say, left tank) has malfunction, the aircraft only will consume fuel from another side (say, right tank). This causes one side to be heavier than the other side, and this asymmetricity about xz plane requires left-wing to create more lift than the right-wing section. The differential aileron deflection has the task for lateral trim. This incidence consequently generates an adverse yaw that requires a directional trim. Due to this process, both aileron and rudder (δ_A and δ_R) are simultaneously used in the lateral trim.

In a fixed-wing aircraft, the wing (through aileron) is mainly responsible to maintain lateral trim and make the summation of moment to be zero, by generating a necessary left/right wing differential lift.

When the summation of all forces in y direction is zero; and the summation of all moments including aerodynamic rolling moment about x axis is zero, the aircraft is said to have the lateral trim.

$$\sum F_y = 0 \tag{2.44}$$

$$\sum L_{cg} = 0 \tag{2.45}$$

The aerodynamic rolling moment is:

$$L = \frac{1}{2}\rho V^2 SbC_l \tag{1.20}$$

where C_l is the aircraft rolling moment coefficient. The aerodynamic rolling moment coefficient (C_l) is a function of a number of flight and control surface parameters. Using Taylor series expansion, it can be expressed as a linear function of flight variables and control surfaces. By employing short-hand expression and non-dimensionalization technique, this equation is presented as:

$$C_l = C_{l_o} + C_{l_\beta}\beta + C_{l_{\delta A}}\delta_A + C_{l_{\delta R}}\delta_R \tag{2.46}$$

where C_{l_β} is a non-dimensional stability derivative, and $C_{l_{\delta A}}$ and $C_{l_{\delta R}}$ are two non-dimensional control derivatives. All non-dimensional derivatives have the unit of 1/rad.

The rate of change of rolling moment with aileron deflection is a sign of the aileron control effectiveness. It is represented by parameter $C_{l_{\delta_A}}$ which is referred to as the aileron control power derivative, and is estimated [4] by:

$$C_{l_{\delta_A}} = \frac{2C_{L_{\alpha_w}}\tau_a C_r}{Sb}\left[\frac{y^2}{2} + \left(\frac{\lambda - 1}{b/2}\right)\frac{y^3}{3}\right]_{y_1}^{y_2} \tag{2.47}$$

where C_r is the wing root chord, λ is wing taper ratio, τ_a is the aileron effectiveness parameter, and y_1 and y_2 are distances between inboard and outboard of aileron to the fuselage centerline.

The wing taper ratio (λ) is ratio between wing tip chord (C_t) to the wing root chord (C_r):

$$\lambda = \frac{C_t}{C_r} \tag{2.48}$$

Most notable laterally trimmed motion is a coordinated turn, at which the aircraft follows a pure circular path. To achieve a level coordinated turn, all three control surfaces must be employed: (1) Aileron to roll and create bank angle, (2) Rudder to provide an aerodynamic side force to make the turn coordinated, and (3) Elevator to keep the altitude constant. In a coordinated turn, sum of the component of forces along body-fixed y-axis is zero, so that there will be no skilling and no slipping. Hence, the radius of turn and rate of turn are both constant, and we have the steady state condition $\dot{V} = 0$. In another word, there will be no lateral acceleration acts on the aircraft.

2.5   Lateral-Directional Trim

There are multiple flight operations and cases where the aircraft needs to be laterally-directionally trim simultaneously. A few examples were briefly introduced earlier in this chapter. Now, let's consider the combination of lateral trim and directional trim for one case; an OEI condition. We desire to laterally-directionally trim this aircraft while keeping a desired heading. This objective requires no net force along y-axis, and no rolling and yawing moments a bout aircraft cg. Assuming $C_{l_o} = C_{n_o} = C_{y_o} = 0$, we obtain:

$$\sum F_y = 0 \Rightarrow \frac{1}{2}\rho V^2 S\left(C_{y_\beta}\beta + C_{y_{\delta_A}}\delta_A + C_{y_{\delta_R}}\delta_R\right) + W\sin(\phi) + T_y = 0 \tag{2.49}$$

$$\sum L_{cg} = 0 \Rightarrow \frac{1}{2}\rho V^2 Sb\left(C_{l_\beta}\beta + C_{l_{\delta_A}}\delta_A + C_{l_{\delta_R}}\delta_R\right) + L_T = 0 \tag{2.50}$$

$$\sum N_{cg} = 0 \Rightarrow \frac{1}{2}\rho V^2 Sb\left(C_{n_\beta}\beta + C_{n_{\delta_A}}\delta_A + C_{n_{\delta_R}}\delta_R\right) + k_{OEI}N_T = 0 \tag{2.51}$$

where T_y is the component of the thrust in y-direction, L_y is the rolling moment of the thrust about x-axis, and N_y is the yawing moment of the thrust about z-axis. If the engine thrust is along fuselage center line, $T_y = 0$. The rolling and yawing moments of the thrust are:

$$L_T = \sum T_{asym}y_T\sin(i_T) \tag{2.52}$$

$$N_T = \sum T_{asym}y_T \tag{2.53}$$

where y_T is the distance between each engine thrust line and the aircraft cg along y-axis, and i_T is the engine installation angle in the xz plane. The parameter k_{OEI} is introduced in Chap. 6 (Sect. 6.3.4). Three possible unknows in Eqs. (2.49)–(2.51) are β, δ_A and δ_R; these equations may be reformatted into a matrix form as:

$$\begin{bmatrix} C_{y_\beta} & C_{y_{\delta_A}} & C_{y_{\delta_R}} \\ C_{l_\beta} & C_{l_A} & C_{l_{\delta_R}} \\ C_{n_\beta} & C_{n_{\delta_A}} & C_{n_{\delta_R}} \end{bmatrix}\begin{bmatrix} \beta \\ \delta_A \\ \delta_R \end{bmatrix} = \begin{bmatrix} -\frac{W\sin(\phi)+T_y}{\frac{1}{2}\rho U^2 S} \\ -\frac{L_T}{\frac{1}{2}\rho U^2 Sb} \\ -\frac{k_{OEI}N_T}{\frac{1}{2}\rho U^2 Sb} \end{bmatrix} \tag{2.54}$$

These equations can be solved by using Cramer's rule. Note that, the aircraft is in straight level flight with steady bank angle (ϕ), refer to Chap. 6, Sect. 6.3.4 for a method to determine bank angle.

Example 2.3 The right engine of a twin jet transport aircraft with a mass of 40,000 kg is inoperative while flying with a speed of 400 knot at 40,000 ft altitude ($\rho = 0.000587$ slug/ft^3). The aircraft has the following geometry, thrust, and lateral-directional stability and control derivatives:

$$S = 105\ m^2, b = 28\ m, T = 2 \times 65\ kN, i_T = 5\ deg, k_{OEI} = 1.15, y_T = 4.3\ m$$

$$C_{y_\beta} = -0.73\ \frac{1}{rad};\ C_{l_\beta} = -0.11\ \frac{1}{rad};\ C_{n_\beta} = 0.13\ \frac{1}{rad};\ C_{y_{\delta_r}} = 0.14\ \frac{1}{rad};\ C_{l_{\delta_r}} = 0.02\ \frac{1}{rad}$$

$$C_{n_{\delta_r}} = -0.07\ \frac{1}{rad};\ C_{y_{\delta_a}} = 0;\ C_{l_{\delta_a}} = 0.18\ \frac{1}{rad};\ C_{n_{\delta_a}} = -0.02\ \frac{1}{rad}$$

The pilot need to laterally-directionally trim the aircraft, while maintain the level flight. Assume $\phi = 0$.

a. Determine the required rudder and aileron deflections to maintain lateral-directional trim.
b. How much is the sideslip angle?

Solution
The rolling and yawing moments of the thrust:

$$L_T = \sum T_{asym} y_T \sin(i_T) = 65{,}000 \times 4.3 \times \sin(5) = 24{,}360 \,\text{N} \tag{2.52}$$

$$N_T = \sum T_{asym} y_T = 65{,}000 \times 4.3 = 279{,}500 \,\text{N} \tag{2.53}$$

Three unknowns: β, δ_A and δ_R.

$$\begin{bmatrix} C_{y\beta} & C_{y\delta_A} & C_{y\delta_R} \\ C_{l\beta} & C_{l_A} & C_{l\delta_R} \\ C_{n\beta} & C_{n\delta_A} & C_{n\delta_R} \end{bmatrix} \begin{bmatrix} \beta \\ \delta_A \\ \delta_R \end{bmatrix} = \begin{bmatrix} -\frac{W\sin(\phi)+T_y}{\frac{1}{2}\rho U^2 S} \\ -\frac{L_T}{\frac{1}{2}\rho U^2 Sb} \\ -\frac{k_{OEI} N_T}{\frac{1}{2}\rho U^2 Sb} \end{bmatrix} \tag{2.54}$$

$$\begin{bmatrix} -0.73 & 0 & 0.14 \\ -0.11 & 0.18 & 0.02 \\ 0.13 & -0.02 & -0.07 \end{bmatrix} \begin{bmatrix} \beta \\ \delta_A \\ \delta_R \end{bmatrix} = \begin{bmatrix} 0 \\ -\frac{24{,}360}{\frac{1}{2}\times 0.000587 \times (400\times 0.514)^2 \times 105 \times 28} \\ -\frac{1.15 \times 279{,}500}{\frac{1}{2}\times 0.000587 \times (400\times 0.514)^2 \times 105 \times 28} \end{bmatrix} = \begin{bmatrix} 0 \\ -0.001 \\ -0.017 \end{bmatrix}$$

Simultaneous solution of this set of linear equations, using Cramer's rule, we obtain the following solutions:

$$\beta = 0.073 \,\text{rad} = 4.2 \,\text{deg}; \quad \delta_A = -0.005 \,\text{rad} = -0.28 \,\text{deg}; \quad \delta_R = 0.381 \,\text{rad} = 21.8 \,\text{deg}.$$

Example 2.4 The twin engine aircraft shown in Fig. 2.8 has lost its left engine. What control surfaces must be deflected and in what directions; in order to maintain the aircraft trim (longitudinal-lateral-directional) for a level flight, and why?

Solution
This is a full OEI problem, that we need to also discuss longitudinal trim. The solution is provided in five steps:

a. **Rudder deflection (δ_R):** In order to trim the aircraft directionally, a *negative* rudder (to the right) must be applied. This will produce a negative side force and a positive yawing moment that opposes the negative yawing moment of asymmetric thrust of right operating engine.

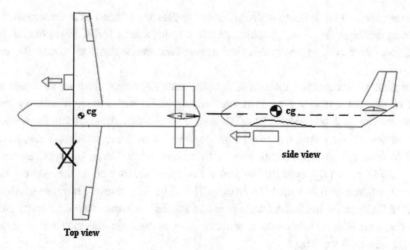

Fig. 2.8 Longitudinal-lateral-directional trim in an OEI condition

b. **Aileron deflection (δ_A):** Sine the vertical tail (rudder) is above the aircraft cg, a negative rudder deflection will produce a negative rolling moment. In order to trim the aircraft laterally, a **positive** aileron (right aileron up and left aileron down) must be applied. This will produce a positive rolling moment that opposes the negative rolling moment of rudder deflection.

c. **Elevator deflection (δ_E):** Since the engine thrust is below the aircraft cg, a positive pitching moment will be created. This will be trimmed (longitudinally) by producing a negative pitching moment through applying a positive elevator (down).

d. However, we assumed that the aircraft was originally trimmed before OEI condition. This means that the loss of one engine will decrease the original positive pitching moment of both engines. This means the original negative pitching moment of the elevator deflection must be reduced.

e. But since the aircraft is losing one engine, the aircraft speed will be decreased. This also means that the original negative moment of elevator deflection is decreased too. In this case, the amount of elevator deflection is **based** on the aircraft configuration. Therefore, the final elevator deflection is the summation of all three elevator deflections out of three cases.

2.6   Trim Tab

In a fixed-wing aircraft, a pilot needs to deflect a control surface (e.g., elevator) to apply a control moment to provide trim. In order to deflect elevator and aileron, stick/yoke needs to be deflected, while rudder pedals are pushed to deflect rudder. Pushing/pulling/deflecting stick and pushing pedals require a pilot force. Since human force is limited, aircraft designers are employing means and mechanisms to reduce the required pilot force.

Moreover, in unexpected incidents, it is necessary for providing an alternate manual control of control surfaces even in a large transport aircraft with hydraulic system. For instance, in the past history flight of Boeing 747 (British Airways on 24 June 1982), there is at least three cases where all four engines become inoperative. Thus, pilot must be able to trim the jumbo aircraft with his/her body force. Large transport aircraft such as Boeing 737 are equipped with horizontal tail trim tab, as well as the vertical tail trim tab. Some modern airliners such as Boeing 787 (Fig. 2.9) doesn't have an elevator trim tab. Instead, the entire horizontal stabilizer is deflected. A trimmable horizontal stabilizer allows for a smaller elevator which reduces drag and allows the elevator to streamline with the horizontal stabilizer.

One method to reduce stick force is to utilize *trim tab* by which the force can be reduced even to zero. *Trim tabs* are frequently used in reversible flight control systems (e.g., mechanical). However, trim tabs are utilized even in very large transport aircraft; due to the fact that, the loss of all engines are conceivable; thus, pilot must be able to trim the jumbo aircraft with his/her body force. In general, *trim tab* serve two functions: (1) it provides the ability to zero-out the stick/wheel/yoke force, and (2) it provides aircraft speed stability at the trim speed. An area of the tail is dedicated to aid the elevator system in maintaining longitudinal trim.

Trim tabs are used to reduce the force the pilot applies to the stick to zero. Tab ensures that the pilot will not tire for holding the stick/yoke/wheel in a prolonged flight. Trailing edge tabs are employed as variable trimming devices, operated by stick/wheel directly from the cockpit.

Control surfaces may be aerodynamically balanced by employing a tab located at the trailing edge of the control surface. Tabs are secondary control surfaces placed at the trailing edges of the primary control surfaces. Figure 2.10 depicts elevator tab of horizontal tail and rudder tab of vertical tail.

Fig. 2.9 Airliner Boeing 787

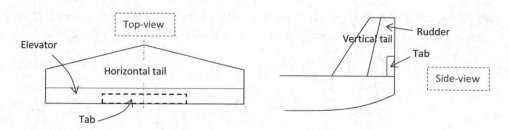

Fig. 2.10 Tail, control surface, and tab

There are a number of tabs used in various aircraft in order to considerably reduce the hinge moment and control force. The most basic tab is a *trim tab*; as the name implies, it is used on elevators to longitudinally trim the aircraft in a cruising flight.

There are two hinges in a lifting surface (Fig. 2.11) when its control surface possesses a tab: (1) one for control surface deflection (δ_E), and (2) one for tab deflection (δ_t). Trim tabs may be adjusted when the aircraft is on the ground; or may be manually operated and set by the pilot during flight. To achieve a zero cockpit control force, the trim tab is deflected opposite to the elevator deflection. The tab aerodynamic force (L_t) is on the opposite direction to the elevator aerodynamic force (L_E). Hence, if the correct tab deflection is applied, the elevator hinge moment (here, clockwise) can be aerodynamically balanced by an opposite tab hinge moment (here, counter-clockwise). When elevator deflection is positive ($+\delta_E$), the tab deflection is negative ($-\delta_t$), and vice versa.

Since, the moment arm of the tab (x_t) is greater than the moment arm of the elevator (x_E), the lift force of the tab portion can be smaller than the lift of the elevator portion. This requires a smaller tab size compared with the elevator size. The tab-to-control-surface-chord ratio is usually about 0.2–0.4. A similar discussion holds true for the rudder and vertical tail.

When the elevator hinge moment is balanced, the sum of two moments shall be zero:

$$\sum H = 0 \rightarrow H_E = H_t \tag{2.55}$$

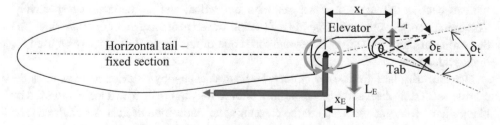

Fig. 2.11 Horizontal tail, elevator, and trim tab mechanism

where H_E stands for elevator hinge moment, and H_t for elevator hinge moment. Each hinge moment is equal to an aerodynamic force multiplied by its arm to the hinge.

$$H_E = L_E x_E \tag{2.56}$$

$$H_t = L_t x_t \tag{2.57}$$

The tab arm (x_t) is the distance between the center of lift tab to the elevator hinge (not the tab hinge). It is not easy for determine the lift center of tab and elevator. The entire tail has one aerodynamic force (tail lift, L_h), but each of the three segments (1. fixed part of the tail, 2. Elevator/rudder, and 3. tab) have a specific share.

$$L_h = L_{h-fixed} + L_E + L_t \tag{2.58}$$

A comprehensive aerodynamic analysis [5] or a CFD simulation for tail is required to determine the aerodynamic force of each segment.

When a tail is equipped with a tab, the elevator *hinge moment* is aerodynamically balanced. The hinge moment created by a control surface is defined similar to other aircraft aerodynamic moments as:

$$H = \frac{1}{2} \rho U_1^2 S_E C_E C_h \tag{2.59}$$

where S_E denotes the planform area of the elevator, and C_E denotes the mean aerodynamic chord of the elevator. The parameter C_h is the hinge moment coefficient and is given by:

$$C_h = C_{h_o} + C_{h_\alpha} \alpha_h + C_{h_{\delta_E}} \delta_E + C_{h_{\delta_t}} \delta_t \tag{2.60}$$

where α_h is the angle of attack of the tail, and δ_t is the tab deflection. The tab deflection is proportional to the geometry of the tab, tab arm, and control surface deflection. The parameter C_{ho} is the hinge moment coefficient for $\alpha_c = \delta_c = 0$; it is zero for a symmetrical airfoil. The parameters C_{h_α}, and $C_{h_{\delta_E}}$, and $C_{h_{\delta_t}}$ are three non-dimensional derivatives; they are introduced in Chap. 5.

These three derivatives are the partial derivatives of hinge moment coefficient (C_h) with respect to tail angle of attack, elevator deflection, and tab deflection respectively. The effectiveness of the tab ($C_{h_{\delta_t}}$) is a function of tab geometry and tab hinge line. The derivation of these derivatives is beyond the scope of this book; Ref. [2] may be used for their evaluations.

Elevator trim tabs are usually deflected—from cockpit—by a device referred to as the *trim wheel*. Trim wheel and trim tab assist a pilot to longitudinally trim a large aircraft with his/her hand force, and to keep a large elevator at any deflection needed at any speed. The combination of tail, elevator and tab is sometimes referred to as the "trimmable horizontal stabilizer". Trim tabs are mainly used during cruising flight, and not during the takeoff rotation of the aircraft. Although the trim tab has a great impact on the hinge moment, it has only a slight effect on the tail lift.

When a tab is employed to aerodynamically balance the control surface, pilot can leave the stick, so, the aircraft is in *stick-free* mode. Tab can aerodynamically hold the elevator (consequently stick) on the desired location. However, when an aircraft has no tab for a tail, the control surface cannot be aerodynamically balanced. In such condition, the aircraft is in *stick-fixed* mode, since the pilot has to hold the control column (i.e., stick) by hand.

When an aircraft is longitudinally trimmed, while the stick is left free, two moments should be nullified: (1) Aircraft aerodynamic pitching moment (M), (2) Elevator hinge moment (H). Thus, the governing equations for a stick-free trim flight are:

$$M = 0 \Rightarrow C_m = C_{m_o} + C_{m_\alpha}\alpha + C_{m_{\delta_E}}\delta_E = 0 \tag{2.61}$$

$$H = 0 \Rightarrow C_h = C_{h_o} + C_{h_{\alpha_h}}\alpha_h + C_{h_{\delta_E}}\delta_E + C_{h_{\delta_t}}\delta_t = 0 \tag{2.62}$$

Two unknows in this set of equations are elevator deflection and tab deflection (δ_E, δ_t). They are determined by simultaneous solution of these linear equations.

2.7 Problems

1. A large transport aircraft with a takeoff weight of 150,000 lb, and a wing area of 1800 ft^2 is cruising with the velocity of 460 knot at 30,000 ft. The aircraft has the following stability and control derivatives:

$$C_{m_\alpha} = -1.1 \frac{1}{\text{rad}}, C_{L_\alpha} = 5.5 \frac{1}{\text{rad}}, C_{L_{\delta_E}} = 0.3 \frac{1}{\text{rad}}, C_{m_{\delta_E}} = -0.72 \frac{1}{\text{rad}},$$
$$C_{L_o} = 0.12, C_{mo} = -0.06$$

   Determine the elevator deflection (δ_E) and aircraft angle of attack for this trimmed flight condition.

2. A large transport aircraft with a takeoff weight of 120,000 lb, and a wing area of 1600 ft^2 is cruising with the velocity of 400 knot at 40,000 ft. The aircraft has the following stability and control derivatives:

$$C_{m_\alpha} = -1.8 \frac{1}{\text{rad}}, C_{L_\alpha} = 5.7 \frac{1}{\text{rad}}, C_{L_{\delta_E}} = 0.4 \frac{1}{\text{rad}}, C_{m_{\delta_E}} = -0.9 \frac{1}{\text{rad}},$$
$$C_{L_o} = 0.2, C_{mo} = -0.04$$

Determine the elevator deflection (δ_E) and aircraft angle of attack for this trimmed flight condition.

3. A large transport aircraft with a maximum take-off weight of 160,000 lb is equipped with two turbofan engines each generating 28,000 lb of thrust. Right after takeoff—when airspeed is 110 knot—right engine becomes inoperative. The lateral distance between two engines is 45 ft. Other characteristics of the aircraft are as follows:

$$C_{L_{\alpha_V}} = 4.6 \frac{1}{\text{rad}}; S = 1350 \, \text{ft}^2; b = 110 \, \text{ft}; S_V = 260 \, \text{ft}^2;$$

$$l_v = 70 \, \text{ft}, \tau_r = 0.4, \eta_v = 0.95$$

Determine how much rudder deflection is needed (and to which direction) to directionally trim the aircraft.

4. A large transport aircraft with a maximum take-off weight of 200,000 lb is equipped with two turbofan engines each generating 36,000 lb of thrust. Right after takeoff—when airspeed is 1150 knot—left engine becomes inoperative. The lateral distance between two engines is 50 ft. Other characteristics of the aircraft are as follows:

$$C_{L_{\alpha_V}} = 5.1 \frac{1}{\text{rad}}; S = 1500 \, \text{ft}^2; b = 130 \, \text{ft}; S_V = 280 \, \text{ft}^2;$$

$$l_v = 76 \, \text{ft}, \tau_r = 0.5, \eta_v = 0.96$$

Determine how much rudder deflection is needed (and to which direction) to directionally trim the aircraft.

5. The right engine of a twin jet transport aircraft with a mass of 50,000 kg is inoperative while flying with a speed of 300 knot at 30,000 ft altitude. The aircraft has the following geometry, thrust, and lateral-directional stability and control derivatives:

$$S = 140 \, \text{m}^2, b = 34 \, \text{m}, T = 2 \times 80 \, \text{kN}, i_T = 4 \, \text{deg}, k_{OEI} = 1.15, y_T = 5.2 \, \text{m}$$

$$C_{y_\beta} = -0.6 \frac{1}{\text{rad}}; C_{l_\beta} = -0.2 \frac{1}{\text{rad}}; C_{n_\beta} = 0.3 \frac{1}{\text{rad}};$$

$$C_{y_{\delta_r}} = 0.18 \frac{1}{\text{rad}}; C_{l_{\delta_r}} = 0.03 \frac{1}{\text{rad}}$$

$$C_{n_{\delta_r}} = -0.08 \frac{1}{\text{rad}}; C_{y_{\delta_a}} = 0; C_{l_{\delta_a}} = 0.22 \frac{1}{\text{rad}}; C_{n_{\delta_a}} = -0.04 \frac{1}{\text{rad}}$$

The pilot needs to laterally-directionally trim the aircraft, while maintain the level flight. Assume $\phi = 0$.

6. The right engine of a twin jet transport aircraft with a mass of 50,000 kg is inoperative while flying with a speed of 300 knot at 30,000 ft altitude. The aircraft has the following geometry, thrust, and lateral-directional stability and control derivatives:

$$S = 110\,m^2, b = 26\,m, T = 2 \times 60\,kN, i_T = 3 \text{ deg}, k_{OEI} = 1.2, y_T = 3.5\,m$$

$$C_{y_\beta} = -1.1\,\frac{1}{rad}; C_{l_\beta} = -0.4\,\frac{1}{rad}; C_{n_\beta} = 0.5\,\frac{1}{rad};$$
$$C_{y_{\delta_r}} = 0.1\,\frac{1}{rad}; C_{l_{\delta_r}} = 0.06\,\frac{1}{rad}$$

$$C_{n_{\delta_r}} = -0.1\,\frac{1}{rad}; C_{y_{\delta_a}} = 0; C_{l_{\delta_a}} = 0.14\,\frac{1}{rad}; C_{n_{\delta_a}} = -0.07\,\frac{1}{rad}$$

The pilot needs to laterally-directionally trim the aircraft, while maintain the level flight. Assume $\phi = 0$.

7. A large transport aircraft (Shown below) is cruising at 20,000 ft altitude with an airspeed of 670 ft/s. Because of a technical problem in middle of the flight, all the fuel contained in the right fuel tanks is suddenly pumped into the left fuel tanks. Assume the total fuel weight in the left fuel tanks is 10% of the aircraft weight and the right fuel tanks are empty. The average distance between fuel cg to the fuselage centerline is 25 ft. Determine how much ailerons (and in what direction) must be deflected in order to maintain the lateral trim. Ignore directional moment and sideslip force created due to this deflection.

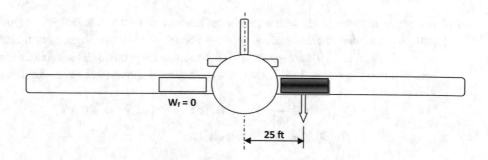

$$S = 600,000\,lb; S = 5500\,ft^2, b = 196\,ft, C_{l_{\delta A}} = 0.013\,\frac{1}{rad}$$

8. Assume a Cessna 182 pilot in a climbing flight sets the throttle at 60% and elevator setting at $-1.2°$ (Sea level). Assume the engine thrust line is 0.5 in above the aircraft center of gravity. Determine the trim conditions of the aircraft (i.e., what are aircraft speed, angle of attack, and climb angle?).

$$W = 3100\,lb, S = 174\,ft^2, C = 4.5\,ft, P = 230\,hp, \eta_P = 0.8$$

$$C_{m_o} = 0.04; C_{D_o} = 0.028; C_{L_o} = 0.3; C_{L_\alpha} = 5.1\,\frac{1}{rad};$$

$$C_{m_\alpha} = -1.9 \frac{1}{\text{rad}}; C_{L_{\delta_E}} = 0.48 \frac{1}{\text{rad}}$$

$$C_{m_{\delta_E}} = -1.4 \frac{1}{\text{rad}}; C_{D_{\delta_E}} = 0; C_{D_\alpha} = 0.06 \frac{1}{\text{rad}}$$

9. The following aircraft is cruising with a speed of 120 knot at 20,000 ft altitude. How much tail lift must be produced, so that the aircraft maintains longitudinal trim? Assume: (1) aircraft angle of attack is zero; and (2) ignore the pitching moment of wing and horizontal tail.

$$\text{m} = 18,000\,\text{kg}, \text{Cm}_{\text{ac}} = -0.02, \text{S} = 90\,\text{m}^2, \text{MAC} = 2.6\,\text{m},$$

$$\text{x}_{\text{cg}} = 16\,\text{m}, \text{x}_{\text{ac}} = 15.5\,\text{m}, \text{x}_{\text{ach}} = 27\,\text{m}$$

10. A business jet aircraft is cruising with an airspeed of 200 knot and an angle of attack of 3°. The aircraft is equipped with elevator trim tab. What elevator deflection and tab deflection are required, in order to leave the stick free?

$$\alpha_h = -1.6\,\text{deg}; C_{m_o} = -0.05; C_{m_\alpha} = -2.3 \frac{1}{\text{rad}}; C_{m_{\delta_E}} = -1.2 \frac{1}{\text{rad}}; C_{h_o} = 0;$$

$$C_{h_{\alpha_h}} = -0.2 \frac{1}{\text{rad}}; C_{h_{\delta_E}} = -1.6 \frac{1}{\text{rad}}; C_{h_{\delta_t}} = 3.5 \frac{1}{\text{rad}}$$

11. A large transport aircraft with a maximum take-off weight of 120,000 lb is equipped with two turbofan engines each generating 25,000 lb of thrust. Right after takeoff—when airspeed is 110 knot—left engine becomes inoperative. The lateral distance between two engines is 34 ft. Other characteristics of the aircraft are as follows:

$$C_{L_{\alpha_V}} = 4.7 \frac{1}{\text{rad}}; \text{S} = 1100\,\text{ft}^2; \text{b} = 94\,\text{ft}; \text{S}_V = 220\,\text{ft}^2;$$

$$l_v = 60\,\text{ft}, \tau_r = 0.4, \eta_v = 0.97$$

Determine how much rudder deflection is needed (and to which direction) to directionally trim the aircraft.

12. A business jet aircraft with a maximum take-off weight of 10,000 lb is equipped with two turbofan engines each generating 4000 lb of thrust. Right after takeoff—when airspeed is 95 knot—left engine becomes inoperative. The lateral distance between two engines is 12 ft. Other characteristics of the aircraft are as follows:

$$C_{L_{\alpha_V}} = 4.7 \frac{1}{\text{rad}}; \text{S} = 240\,\text{ft}^2; \text{b} = 48\,\text{ft}; \text{S}_V = 56\,\text{ft}^2;$$

$$l_v = 20\,\text{ft}, \tau_r = 0.45, \eta_v = 0.96$$

Determine how much rudder deflection is needed (and to which direction) to directionally trim the aircraft.

References

1. Elgersma, M. R. and B. G. Morton, Nonlinear six-degree-of-freedom aircraft trim, *Journal of Guidance, Control, and Dynamics*, 2000, 23(2); 305–311
2. Roskam J., Airplane flight dynamics and automatic flight controls, 2007, DARCO
3. MIL-STD-1797, Flying Qualities of Piloted Aircraft, Department of Defense, Washington DC, 1997
4. Nelson R., Flight Stability and Automatic Control, McGraw Hill, 1989
5. Anderson J. D., Fundamentals of Aerodynamics, McGraw-Hill, Sixth edition, 2016

Longitudinal Stability

3

3.1 Fundamentals

3.1.1 Definitions

One of the most important characteristics of the dynamic behavior of an aircraft is absolute stability—that is, whether the aircraft is stable or unstable. Stability refers to the tendency of an object (here, aircraft) to oppose any disturbance, and to return to its equilibrium state of motion, if disturbed. Stability implies that the forces acting on the aircraft (thrust, weight, and aerodynamic forces) are in initial directions, and will gain new values, that tend to restore the aircraft to its original equilibrium conditions after it has been disturbed (by a wind gust or other forces). The aircraft stability is inherent, and primarily provided by the air vehicle configuration (without pilot or autopilot interference). Inherent stability is not a function of initial trim conditions, it is a function of aircraft parameters such as geometry, weight, and center of gravity location.

With stability, an increasing disturbance in the state of a dynamic system generates increasing restoring moments/forces. However, with instability, an increasing disturbance in the state generates increasing disturbing moments/forces that can end up into a dynamic system failure. An aircraft is unstable if the flight output diverges without bound from its equilibrium state when the aircraft is impacted by a disturbance. In practice, the flight output will increase to a certain extent, since all flight parameters are limited by nature or physical/mechanical "stops," or the aircraft may crash or break down after the output exceeds a certain magnitude.

From stability point of view, aircraft may have one of the three cases: (1) Stable, (2) Unstable, and (3) Neutrally or critically stable. A neutrally stable aircraft is the one that, if disturbed, will not return to its trim point, and will not move away from trim point, without bound. However, it will depart from initial trim point to a new trim point.

© The Author(s), under exclusive license to Springer Nature Switzerland AG 2022
M. H. Sadraey, *Flight Stability and Control*, Synthesis Lectures on Mechanical
Engineering, https://doi.org/10.1007/978-3-031-18765-0_3

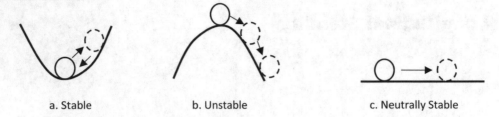

a. Stable   b. Unstable   c. Neutrally Stable

Fig. 3.1 Three cases for stability of a dynamic system

Figure 3.1 illustrate the three cases for a ball in three different conditions. A ball inside a bowl (Fig. 3.1a) creates a stable dynamic system, since if disturbed (i.e., when it is displaced from its equilibrium bottom point), it will return back to bottom of the bowl, due to its weight and shape of the ball.

A ball placed on top of a bowl (Fig. 3.1b) generates an unstable dynamic system, since if disturbed, it cannot return back to the top of the bowl, due to its weight and inverted bowl. Finally, when a ball is placed on a flat surface (Fig. 3.1c), it will create a neutrally stable dynamic system. When it is disturbed (i.e., when it is displaced from its current equilibrium stationary point), it will move to a new location, due to friction between ball and the surface.

Stability is categorized into two modes and can be treated in three axes independently. There are two modes for stability: (1) Static stability, and (2) Dynamic stability. Static stability refers to the tendency of an object (here, aircraft) to oppose any disturbance (by developing forces and moments), while dynamic stability is to return to the equilibrium state of motion, if disturbed.

In general, fixed-wing air vehicles are often stable (e.g., Boeing 767, and AeroVironment RQ-11 Raven), while VTOL aircraft including helicopters and quadcopters (e.g., MD Helicopter 520N and DJI Phantom) and fixed-wing aircraft with VTOL capabilities (e.g., Bell Boeing V-22 Osprey) are inherently unstable. The VTOL aircraft have more number of safety incidents, and the stability is often artificially provided by the automatic flight control systems. When the air vehicle configuration is changed, its stability characteristics will vary.

For military and highly maneuverable aircraft (e.g., General Dynamics F-16 Fighting Falcon and Eurofighter), instability can be acceptable and even desirable. This is illustrated by a bicycle, which has only a small tolerance of disturbance before it will fall over unless the rider shifts his/her weight slightly to correct the effects of the disturbance. Instability often coincides with very high maneuverability and can be desirable in some situations. However, instability requires a control system to correct any small disturbances before the disturbing feedback forces become uncontrollable.

If the air vehicle is not statically stable, the smallest disturbance will cause ever-increasing deviations from the original flight state. A statically stable aircraft will have the "tendency" to return to its original position after a disturbance, but it may overshoot, turn around, go in the opposite direction, overshoot again, and eventually oscillate to destruction. In this case, the aircraft would be statically stable but dynamically unstable. If the oscillations are damped and eventually die out, then the air vehicle is said to be dynamically stable.

In terms of axis, we have: (1) Lateral stability (about x axis), (2) Longitudinal stability (about y axis), and (3) Directional stability (about z axis). An air vehicle is desired to be stable in both modes and about all three axes (x, y, and z). The pitch axis is most critical, and stability about it is called longitudinal stability. Some instability can be tolerated about the roll and yaw axis, which are combined in most analysis and called lateral stability. The concept of stability in two modes and three axes are introduced in the following sections.

The required degree of stability is governed by the flying qualities and the desired comfort level for attendants and airframe tolerance. The desired degree of static stability is expressed in terms of opposition speed to any disturbance. The required degree of dynamic stability is usually specified by the time it takes the motion to damp to half of its initial amplitude.

Longitudinal motion is defined as any flying operation within xz plane. Longitudinal stability is defined as the tendency of an aircraft to oppose any longitudinal disturbance (e.g., up/down gust), and to return to its equilibrium longitudinal state of motion, if disturbed. An aircraft designer should consider the longitudinal stability requirements, and shall make sure that, the aircraft is longitudinally stable within the design flight envelope.

In this chapter, static and dynamic longitudinal stability of an aircraft are examined, and their requirements and analysis techniques are presented. Moreover, the aircraft reactions to longitudinal perturbations and aircraft components contributions to longitudinal stability are discussed. Before analysis techniques are discussed, we need to derive longitudinal motions governing equations.

3.1.2 Dynamic Stability Analysis of a Dynamic System

Dynamics of many physical systems can be modeled by a second order differential equations (in time domain) or a second-order proper transfer function (in s-domain).

In general, the response of a first order system—to a step input—follows a logarithmic path, but for a second order systems, it will be oscillatory. Since the second-order differential equation plays an important role in aircraft dynamics, the properties of a second-order differential equation are further examined. The transfer function of a dynamic system which is represented by a second order differential equation:

$$\frac{d^2x}{dt^2} + a\frac{dx}{dt} + bx = N\dot{f}(t) \tag{3.1}$$

is obtained by applying Laplace transform. Thus:

$$s^2X(s) + aX(s) + bX(s) = N\dot{F}(s) \tag{3.2}$$

Thus, the transfer function is obtained as:

$$\frac{X(s)}{F(s)} = \frac{N}{s^2 + as + b} = \frac{N}{s^2 + 2\zeta\omega_n s + \omega_n^2} \tag{3.3}$$

The *characteristic equation* of this system (i.e., denominator of transfer function) is:

$$s^2 + 2\zeta\omega_n s + \omega_n^2 = 0 \tag{3.4}$$

This second order characteristic equation has two roots. If both roots (s_1 and s_2) have negative real parts, the system is said to be *dynamically stable*. If there is at least one root with a positive real part, the system is *dynamically unstable*. However, if at least real part of one root is positive, and all other roots have negative real parts, the system is *dynamically neutrally stable*.

When the roots of characteristic equation of a second order system are complex conjugate, the behavior of the system can be expressed in terms of three oscillatory motion variables: (1) Damping ratio (ζ), (2) Natural frequency (ω_n), and (3) Period (T).

The roots of this second order characteristic equation are:

$$s_{1,2} = -\zeta\omega_n \pm i\omega_n\sqrt{1 - \zeta^2} \tag{3.5}$$

Any complex number has two parts: (1) Real part (η), (2) Imaginary part (ω)

$$s_{1,2} = \eta \pm i\omega \tag{3.6}$$

The real part is:

$$\eta = -\zeta\omega_n \tag{3.7}$$

The imaginary part is:

$$\omega = \omega_n\sqrt{1 - \zeta^2} \tag{3.8}$$

The real part of the root ($-\zeta\omega_n$) governs the damping of the response and the imaginary part ($\omega_n\sqrt{1 - \zeta^2}$) is the damped natural frequency. When the coefficients of the characteristic equation are known, one can determine the damped natural frequency and damping ratio via Eqs. (3.7) and (3.8).

Since a complete oscillation (full circle) is equivalent to 2π radian, the period of oscillation (T) of any oscillatory mode is:

$$T = \frac{2\pi}{\omega} \tag{3.9}$$

The required degree of dynamic stability may be specified by the time it takes the oscillation to damp to half of its initial amplitude. To measure the damping power of the vehicle, we determine the amplitude ratio per one period of damped oscillation via:

$$\frac{x_1}{x_2} = e^{2\zeta\pi/\sqrt{1-\zeta^2}} \tag{3.10}$$

From Eq. (3.10), we observe that, the transient response of an oscillatory motion (mode) is primarily characterized by damping ratio.

3.2 Longitudinal Governing Equations of Motion

3.2.1 Governing Differential Equations of Motion

The nonlinear six-degrees of freedom aircraft governing equations of motion are derived in Chap. 1 (Eqs. 1.32–1.37). We will now decouple longitudinal motions from lateral-directional equations and collect the longitudinal and lateral-directional equations separately. Three degrees of freedom features a longitudinal motion: (1) Linear motion along x-axis, (2) Linear motion along z-axis, (3) Rotational motion about y-axis. To analyze longitudinal stability, we need to derive longitudinal transfer functions and/or longitudinal state space model.

For linear motion along x-axis, variables are linear displacement (mainly forward, X), linear speed (u), and linear acceleration ($\dot{u}$). For linear motion along z-axis, variables are linear displacement (mainly altitude, h = $-z$), linear speed (w), and linear acceleration ($\dot{w}$). For angular motion about y-axis, variables are angular displacement (pitch angle, θ), angular speed or pitch rate (q), and angular acceleration ($\dot{q}$).

Two force Eqs. (1.50) and (1.52), and one moment Eq. (1.54) govern the longitudinal motion (in xz plane). The equations are repeated here for convenience:

$$m(\dot{u} - V_1 r - R_1 v + W_1 q + Q_1 w) = f_{A_X} + f_{T_X} - mg\theta \cos \Theta_1 \tag{3.11}$$

$$m(\dot{w} - U_1 q - Q_1 u + V_1 p + P_1 v) = -mg(\theta \cos \Phi_1 \sin \Theta_1 - \phi \sin \Phi_1 \cos \Theta_1) + f_{A_Z} + f_{T_Z} \tag{3.12}$$

$$I_{yy}\dot{q} + (I_{xx} - I_{zz})(P_1 r + R_1 p) + I_{xz}(2P_1 p - 2R_1 r) = m_A + m_T \tag{3.13}$$

To analyze longitudinal stability, we need to isolate longitudinal motions from lateral-directional motions. As discussed in Chap. 1 for motions decoupling, we drop lateral-directional flight parameters (i.e., $p = r = \phi = v = 0$) from longitudinal governing equations of motion. Moreover, a level cruising (steady state symmetric straight line) flight is considered, so $W_1 = Q_1 = \Phi_1 = 0$. Furthermore, it is assumed that engine thrust is along x-axis, and will not be impacted by any longitudinal disturbance (i.e., $m_T = f_{T_Z} = 0$ and $T \cos \alpha = T$). For aerodynamic forces along x and z, and pitching moment, the new symbols X, Z and M are utilized respectively. Thus, the equations are simplified to the following form:

$$\dot{u} = \frac{1}{m}(X + T) - g \cos \Theta_1 \theta \tag{3.14}$$

$$\dot{w} - U_1 q = -g(\sin \Theta_1 \theta) + \frac{Z}{m} \tag{3.15}$$

$$\dot{q} = \frac{M}{I_{yy}} \tag{3.16}$$

Next step is to employ the linearized forces (X, Z) and moment (M), as derived in Chap. 1. Two longitudinal forces and one moment equations (Eqs. 1.65, 1.66, and 1.69) are repeated here for convenience.

$$X = \frac{\partial X}{\partial u}u + \frac{\partial X}{\partial \alpha}\alpha + \frac{\partial X}{\partial \delta_E}\delta_E \tag{3.17}$$

$$Z = \frac{\partial Z}{\partial u}u + \frac{\partial Z}{\partial \alpha}\alpha + \frac{\partial Z}{\partial \dot{\alpha}}\dot{\alpha} + \frac{\partial Z}{\partial q}q + \frac{\partial Z}{\partial \delta_E}\delta_E \tag{3.18}$$

$$M = \frac{\partial M}{\partial u}u + \frac{\partial M}{\partial p}\alpha + \frac{\partial M}{\partial \dot{\alpha}}\dot{\alpha} + \frac{\partial M}{\partial q}q + \frac{\partial M}{\partial \delta_E}\delta_E \tag{3.19}$$

Recall that, for aerodynamic forces in xz plane, $X = -D$, and $Z = -L$. The longitudinal aerodynamic forces (lift and drag) and pitching moment can be expressed as a function of all the motion variables. However, in these linear equations, only the terms that are significant have been retained. To keep the format of Eqs. (3.14)–(3.16), we use new parameters for dimensional derivatives:

$$X_u = \frac{\partial X/\partial u}{m}; \quad X_\alpha = \frac{\partial X/\partial \alpha}{m}; \quad X_{\delta_E} = \frac{\partial X/\partial \delta_E}{m} \tag{3.20}$$

$$Z_u = \frac{\partial Z/\partial u}{m}; \quad Z_\alpha = \frac{\partial Z/\partial \alpha}{m}; \quad Z_{\dot{\alpha}} = \frac{\partial Z/\partial \dot{\alpha}}{m}; \quad Z_q = \frac{\partial Z/\partial q}{m}; \quad Z_{\delta_E} = \frac{\partial Z/\partial \delta_E}{m} \tag{3.21}$$

$$M_u = \frac{\partial M/\partial u}{I_{yy}}; \quad M_\alpha = \frac{\partial M/\partial \alpha}{I_{yy}}; \quad M_{\dot{\alpha}} = \frac{\partial M/\partial \dot{\alpha}}{I_{yy}}; \quad M_q = \frac{\partial M/\partial q}{I_{yy}}; \quad M_{\delta_E} = \frac{\partial M/\partial \delta_E}{I_{yy}}$$

(3.22)

Moreover, along x-axis, there are two forces: (1) Drag ($D = -X$), (2) Engine throttle (T). To keep the uniformity, we will use the thrust as a linear function of throttle setting as:

$$T = \frac{\partial T}{\partial \delta_T} \delta_T = C_{\delta_T} \delta_T$$

(3.23)

and

$$X_{\delta_T} = T_{\delta_T} = \frac{\partial T/\partial \delta_T}{m} = \frac{C_{\delta_T}}{m}$$

(3.24)

Typical values for derivative T_{δ_T} lie in the range of 5–15 ft/s^2 for most non-fighter aircraft. Hence, the linear governing equations for longitudinal motion will be reformatted as:

$$\dot{u} = -g\theta \cos \Theta_1 + X_u u + X_\alpha \alpha + X_{\delta_E} \delta_E + X_{\delta_T} \delta_T$$

(3.25)

$$\dot{w} - U_o q = -g\theta(\sin \Theta_o) + Z_u u + Z_\alpha \alpha + Z_{\dot{\alpha}} \dot{\alpha} + Z_q q + Z_{\delta_E} \delta_E$$

(3.26)

$$\dot{q} = M_u u + M_\alpha \alpha + M_{\dot{\alpha}} \dot{\alpha} + M_q q + M_{\delta_E} \delta_E$$

(3.27)

The dimensional derivatives are determined [1] from the following equations:
Aerodynamic force derivatives along x-axis:

$$X_u = \frac{-(C_{D_u} + 2C_{D_1})\bar{q}_1 S}{mU_o}; \quad X_\alpha = \frac{-(C_{D_\alpha} - 2C_{L_1})\bar{q}_1 S}{m}; \quad X_{\delta_E} = \frac{-C_{D_{\delta_E}}\bar{q}_1 S}{m}$$

(3.28)

Aerodynamic force derivatives along z-axis:

$$Z_u = \frac{-(C_{L_u} + 2C_{L_1})\bar{q}_1 S}{mU_o}; \quad Z_\alpha = \frac{-(C_{L_\alpha} + C_{D_1})\bar{q}_1 S}{m}; \quad Z_{\dot{\alpha}} = \frac{-C_{L_{\dot{\alpha}}}\bar{q}_1 SC}{2mu_o};$$

$$Z_q = \frac{-C_{L_q}\bar{q}_1 SC}{2mu_o}; \quad Z_{\delta_E} = \frac{-C_{L_{\delta_E}}\bar{q}_1 S}{m}$$

(3.29)

Aerodynamic pitching moment derivatives (about y-axis):

$$M_u = \frac{\bar{q}_1 S\bar{c}(C_{m_u} + 2C_{m_1})}{I_{yy}U_o}; \quad M_\alpha = \frac{\bar{q}_1 S\bar{c}C_{m\alpha}}{I_{yy}}; \quad M_{\dot{\alpha}} = \frac{\bar{q}_1 S\bar{c}^2 C_{m\dot{\alpha}}}{2I_{yy}U_o}; \quad M_{\delta_E} = \frac{\bar{q}_1 S\bar{c}C_{m\delta_E}}{I_{yy}}$$

(3.30)

We are interested in observing variations of the following motion variables: (1) Angle of attack (α), (2) Linear airspeed (u), (3) Pitch angle (θ), and (4) Pitch rate (q). Two primary control input variables to longitudinal motions are: (1) Elevator deflection (δ_E), and (2) Engine throttle (δ_T). In a pure longitudinal motion, pitch rate is q $= \dot{\theta}$, so we can write $\dot{q} = \ddot{\theta}$. Furthermore, as the angle of attack (α) is defined in Chap. 1, $w =$ U$_0$ α, so we can write, $\dot{w} = U_o\dot{\alpha}$. Hence, Eqs. (3.15)–(3.18) will be rewritten as:

$$\dot{u} = -g\theta \cos \Theta_o + X_u u + X_\alpha \alpha + X_{\delta_E}\delta_E + T_{\delta_T}\delta_T \tag{3.31}$$

$$U_o\dot{\alpha} - U_o\dot{\theta} = -g\theta(\sin \Theta_o) + Z_u u + Z_\alpha \alpha + Z_{\dot{\alpha}}\dot{\alpha} + Z_q q + Z_{\delta_E}\delta_E \tag{3.32}$$

$$\ddot{\theta} = M_u u + M_\alpha \alpha + M_{\dot{\alpha}}\dot{\alpha} + M_q q + M_{\delta_E}\delta_E \tag{3.33}$$

These linear differential equations govern the pure longitudinal motion and can be readily presented into two forms: (1) Transfer functions, (2) State space model. In general, classical control simply employs transfer functions (s-domain), while modern control is often based on state-space (time-domain) models. These longitudinal dynamic models are derived in the following two sections.

3.2.2 Longitudinal Transfer Functions

A transfer function is defined as the Laplace transform of the output over the Laplace transform of the input. It is the ratio of two polynomials in s that represents a dynamic system. A transfer function can only accommodate single-input–single-output (SISO) systems.

To derive longitudinal transfer functions, we first need to apply Laplace transform (L) to all three differential equations (3.31)–(3.33), assuming zero initial conditions. Recall that (L($\dot{x}$(t)) = s X(s)) and (L($\ddot{x}$(t)) = s^2 X(s)).

$$su(s) = -g \cos \Theta_o\theta(s) + X_u u(s) + X_\alpha \alpha(s) + X_{\delta_E}\delta_E(s) + T_{\delta_T}\delta_T(s) \tag{3.34}$$

$$U_o s\alpha(s) - U_o s\theta(s) = -g \sin \Theta_o\theta(s) + Z_u u(s) + Z_\alpha \alpha(s) + Z_{\dot{\alpha}}s\alpha(s) \\ + Z_q s\theta(s) + Z_{\delta_E}\delta_E(s) \tag{3.35}$$

$$s^2\theta(s) = M_u u(s) + M_\alpha \alpha(s) + M_{\dot{\alpha}}s\alpha(s) + M_q s\theta(s) + M_{\delta_E}\delta_E(s) \tag{3.36}$$

Then, these equations are reformatted as factors of two inputs ($\delta_E(s)$, $\delta_T(s)$) and three outputs ($u(s), \alpha(s), \theta(s)$):

$$(s - X_u)u(s) - X_\alpha \alpha(s) + g \cos \Theta_o\theta(s) = X_{\delta_E}\delta_E(s) + T_{\delta_T}\delta_T(s) \tag{3.37}$$

$$-Z_u u(s) + [s(U_o - Z_{\dot{\alpha}}) - Z_\alpha]\alpha(s) + \big[g \sin \Theta_o - s\big(Z_q + U_o\big)\big]\theta(s) = Z_{\delta_E}\delta_E(s)$$
(3.38)

$$-M_u u(s) - [M_{\dot{\alpha}}s + M_\alpha]\alpha(s) + \big(s^2 - M_q s\big)\theta(s) = M_{\delta_E}\delta_E(s) \qquad (3.39)$$

Writing these three equations in matrix format will allows us to use Cramer's rule to derive longitudinal transfer functions. For stability analysis, we are interested only in three transfer functions $\frac{u(s)}{\delta_E(s)}$, $\frac{\alpha(s)}{\delta_E(s)}$, and $\frac{\theta(s)}{\delta_E(s)}$, so we temporality drop throttle input ($\delta_T(s)$) from the equations.

$$
\begin{bmatrix}
(s - X_u) & -X_\alpha & g \cos \Theta_1 \\
-Z_u & s(U_1 - Z_{\dot{\alpha}}) - Z_\alpha & g \sin \Theta_o - s\big(Z_q + U_o\big) \\
-M_u & -[M_{\dot{\alpha}}s + M_\alpha] & \big(s^2 - M_q s\big)
\end{bmatrix}
\begin{bmatrix}
\frac{u(s)}{\delta_E(s)} \\
\frac{\alpha(s)}{\delta_E(s)} \\
\frac{\theta(s)}{\delta_E(s)}
\end{bmatrix}
=
\begin{bmatrix}
X_{\delta_E} \\
Z_{\delta_E} \\
M_{\delta_E}
\end{bmatrix}
\qquad (3.40)
$$

The Cramer's rule is an algebraic formula for the solution of a system of linear algebraic equations (i.e., A x = b). The solution is expressed as the ratio of two determinants ($x_i = \frac{\det(A_i)}{\det(A)}$): (1) Denominator is the determinant of the coefficient matrix (A), (2) Numerator is the coefficient matrix when its relevant column is replaced by the by the column vector of right-hand-side of the equations. For the first unknown (i.e., the airspeed-to-elevator-deflection transfer function; $\frac{u(s)}{\delta_E(s)}$), in the numerator, the first column of the coefficient matrix is replaced by the by the column vector of right-hand-side.

$$
\frac{u(s)}{\delta_E(s)} =
\frac{
\left|
\begin{bmatrix}
X_{\delta_E} & -X_\alpha & g \cos \Theta_1 \\
Z_{\delta_E} & s(U_o - Z_{\dot{\alpha}}) - Z_\alpha & g \sin \Theta_1 - s\big(Z_q + U_o\big) \\
M_{\delta_E} & -[M_{\dot{\alpha}}s + M_\alpha] & \big(s^2 - M_q s\big)
\end{bmatrix}
\right|
}{
\left|
\begin{bmatrix}
(s - X_u) & -X_\alpha & g \cos \Theta_o \\
-Z_u & s(U_o - Z_{\dot{\alpha}}) - Z_\alpha & g \sin \Theta_o - s\big(Z_q + U_o\big) \\
-M_u & -[M_{\dot{\alpha}}s + M_\alpha] & \big(s^2 - M_q s\big)
\end{bmatrix}
\right|
}
\qquad (3.41)
$$

Similarly, for the second unknown (i.e., the angle-of-attack-to-elevator-deflection transfer function; $\frac{\alpha(s)}{\delta_E(s)}$), in the numerator, the second column of the coefficient matrix is replaced by the by the column vector of right-hand-side.

$$
\frac{\alpha(s)}{\delta_E(s)} =
\frac{
\left|
\begin{bmatrix}
(s - X_u) & X_{\delta_E} & g \cos \Theta_o \\
-Z_u & Z_{\delta_E} & g \sin \Theta_o - s\big(Z_q + U_o\big) \\
-M_u & M_{\delta_E} & \big(s^2 - M_q s\big)
\end{bmatrix}
\right|
}{
\left|
\begin{bmatrix}
(s - X_u) & -X_\alpha & g \cos \Theta_o \\
-Z_u & s(U_o - Z_{\dot{\alpha}}) - Z_\alpha & g \sin \Theta_o - s\big(Z_q + U_o\big) \\
-M_u & -[M_{\dot{\alpha}}s + M_\alpha] & \big(s^2 - M_q s\big)
\end{bmatrix}
\right|
}
\qquad (3.42)
$$

Finally, for the third unknown (i.e., the pitch-angle-to-elevator-deflection transfer function; $\frac{\theta(s)}{\delta_E(s)}$), in the numerator, the third column of the coefficient matrix is replaced by the by the column vector of right-hand-side.

$$\frac{\theta(s)}{\delta_E(s)} = \frac{\left\| \begin{bmatrix} (s - X_u) & -X_\alpha & X_{\delta_E} \\ -Z_u & s(U_o - Z_{\dot\alpha}) - Z_\alpha & Z_{\delta_E} \\ -M_u & -[M_{\dot\alpha}s + M_\alpha] & M_{\delta_E} \end{bmatrix} \right\|}{\left\| \begin{bmatrix} (s - X_u) & -X_\alpha & g\cos\Theta_1 \\ -Z_u & s(U_o - Z_{\dot\alpha}) - Z_\alpha & g\sin\Theta_o - s(Z_q + U_o) \\ -M_u & -[M_{\dot\alpha}s + M_\alpha] & (s^2 - M_q s) \end{bmatrix} \right\|} \tag{3.43}$$

Recall from algebra that, determinant of a 3×3 matrix is determined as follows:

$$\left\| \begin{bmatrix} a & b & c \\ d & e & f \\ g & h & i \end{bmatrix} \right\| = a(ei - fh) - b(di - fg) + c(dh - eg) \tag{3.44}$$

By employing this formula and expanding the coefficients, the following transfer functions are obtained. The airspeed-to-elevator-deflection transfer function:

$$\frac{u(s)}{\delta_E(s)} = \frac{A_u s^3 + B_u s^2 + C_u s + D_u}{A_1 s^4 + B_1 s^3 + C_1 s^2 + D_1 s + E_1} \tag{3.45}$$

The angle-of-attack-to-elevator-deflection transfer function:

$$\frac{\alpha(s)}{\delta_E(s)} = \frac{A_\alpha s^3 + B_\alpha s^2 + C_\alpha s + D_\alpha}{A_1 s^4 + B_1 s^3 + C_1 s^2 + D_1 s + E_1} \tag{3.46}$$

The pitch-angle-to-elevator-deflection transfer function:

$$\frac{\theta(s)}{\delta_E(s)} = \frac{A_\theta s^2 + B_\theta s + C_\theta}{A_1 s^4 + B_1 s^3 + C_1 s^2 + D_1 s + E_1} \tag{3.47}$$

Note that, all three transfer functions have the same fourth-order denominator with the following coefficients:

$$A_1 = U_o - Z_{\dot\alpha}$$

$$B_1 = -(U_o - Z_{\dot\alpha})(X_u + M_q) - Z_\alpha - M_{\dot\alpha}(U_o + Z_q)$$

$$C_1 = X_u\{M_q(U_o - Z_{\dot\alpha}) + Z_\alpha + M_{\dot\alpha}(U_o + Z_q)\} + M_q Z_\alpha \\ - Z_u X_\alpha + M_{\dot\alpha}g\sin\Theta_o - M_\alpha(U_o + Z_q)$$

$$D_1 = g \sin \Theta_o (M_\alpha - M_{\dot\alpha} X_u) + g \cos \Theta_o \{Z_u M_{\dot\alpha} + M_u (U_o - Z_{\dot\alpha})\}$$
$$- M_u X_\alpha (U_o + Z_q) + Z_u X_\alpha M_q + X_u \{M_\alpha (U_o + Z_q) - M_q Z_\alpha\}$$

$$E_1 = g \cos \Theta_o \{M_\alpha Z_u - Z_\alpha M_u\} + g \sin \Theta_o (M_u X_\alpha - X_u M_\alpha) \qquad (3.48)$$

The coefficients for airspeed-to-elevator-deflection transfer function are:

$$A_u = X_{\delta_E}(U_o - Z_{\dot\alpha})$$

$$B_u = -X_{\delta_E}\{M_q(U_o - Z_{\dot\alpha}) + Z_\alpha + M_{\dot\alpha}(U_o + Z_q)\} + Z_{\delta_E} X_\alpha$$

$$C_u = X_{\delta_E}\{M_q Z_\alpha + M_{\dot\alpha} g \sin \Theta_o - M_{\dot\alpha}(U_o + Z_q)\} + Z_{\delta_E}\{-M_{\dot\alpha} g \cos \Theta_o - X_\alpha M_q\}$$
$$+ M_{\delta_E}\{X_\alpha(U_o + Z_q) - (U_o - Z_{\dot\alpha})g \cos \Theta_o\}$$

$$D_u = X_{\delta_E} M_\alpha g \sin \Theta_o - Z_{\delta_E} M_\alpha g \cos \Theta_o + M_{\delta_E}(Z_\alpha g \cos \Theta_o - X_\alpha g \sin \Theta_o) \qquad (3.49)$$

The coefficients for angle-of-attack-to-elevator-deflection transfer function are:

$$A_\alpha = Z_{\delta_E}$$

$$B_\alpha = X_{\delta_E} Z_u - Z_{\delta_E}(M_q + X_u) + M_{\delta_E}(U_o + Z_q)$$

$$C_\alpha = X_{\delta_E}\{(U_o + Z_q)M_u - M_q Z_u\} + Z_{\delta_E} M_q X_u$$
$$- M_{\delta_E}\{g \sin \Theta_o + X_u(U_o + Z_q)\}$$

$$D_\alpha = -X_{\delta_E} M_u g \sin \Theta_o + Z_{\delta_E} M_u g \cos \Theta_o$$
$$+ M_{\delta_E}(X_u g \sin \Theta_o - Z_u g \cos \Theta_o) \qquad (3.50)$$

The coefficients for pitch-angle-to-elevator-deflection transfer function are:

$$A_\theta = Z_{\delta_E} M_{\dot\alpha} + M_{\delta_E}(U_o - Z_{\dot\alpha})$$

$$B_\theta = X_{\delta_E}\{Z_u M_{\dot\alpha} + M_u(U_o - Z_{\dot\alpha})\} + Z_{\delta_E}(M_\alpha - M_{\dot\alpha} X_u)$$
$$- M_{\delta_E}\{Z_\alpha + X_u(U_o - Z_{\dot\alpha})\}$$

$$C_\theta = X_{\delta_E}(M_\alpha Z_u - Z_\alpha M_u) + Z_{\delta_E}(X_\alpha M_u - M_\alpha X_u)$$
$$+ M_{\delta_E}(Z_\alpha X_u - X_\alpha Z_u) \qquad (3.51)$$

Derivations of Eqs. (3.48)–(3.51) are left to the reader as a practice problem. As observed, the longitudinal transfer functions are functions of longitudinal stability and control derivatives, as well as the initial flight conditions (e.g., trim airspeed (U_o) and pitch angle (Θ_o)). The applications of these transfer functions will be presented in Sect. 3.4, as well as in Chap. 5 for longitudinal control.

3.2.3 Longitudinal State Space Model

Another form of dynamic formulation of the longitudinal motion is the state space representation, which is a set of linear first-order linear differential equations, and a set of linear algebraic equations. The model is expressed in the following general format:

$$\dot{x} = Ax + Bu$$
$$y = Cx + Du \tag{3.52}$$

where the A, B, C, and D matrices, x denote the state variables, u the input variables, and y the output variables. The set of equations (3.52) is a *state-space* description of a dynamic system, that can accommodate multiple-input–multiple-output (MIMO) systems.

We are interested in state variables to be: (1) Angle of attack (α), (2) Linear airspeed (u), (3) Pitch angle (θ), and (4) Pitch rate (q). Two input (control) variables in longitudinal motions are: (1) Elevator deflection (δ_E), and (2) Engine throttle (δ_T).

Equations (3.31)–(3.33) in term of state variables [u, α, q, θ] and input variables [δ_E and δ_T] are reformatted. For simplicity, we drop derivatives due to rate of change of angel of attack ($\dot{\alpha}$), since they do not often have significant impact on longitudinal stability. Hence, the following state equations are obtained:

$$\dot{u} = \theta(-g\cos\Theta_o) + X_u u + X_\alpha \alpha + X_{\delta_E}\delta_E + T_{\delta_T}\delta_T \tag{3.53}$$

$$\dot{\alpha} = \frac{1}{U_o}\left[-g\theta(\sin\Theta_o) + U_o q + Z_u u + Z_\alpha \alpha + Z_q q + Z_{\delta_E}\delta_E\right] \tag{3.54}$$

$$\dot{q} = M_u u + M_\alpha \alpha + M_q q + M_{\delta_E}\delta_E \tag{3.55}$$

$$\dot{\theta} = q \tag{3.56}$$

Last Eq. (3.56) is just added to demonstrate the relation between pitch angle and pitch rate in a pure longitudinal motion. Equations (3.53)–(3.56) can be expressed into matrix form—state equation:

$$
\begin{bmatrix} \dot{u} \\ \dot{\alpha} \\ \dot{q} \\ \dot{\theta} \end{bmatrix} = \begin{bmatrix} X_u & X_\alpha & 0 & -g\cos\Theta_o \\ \frac{Z_u}{U_o} & \frac{Z_\alpha}{U_o} & 1+\frac{Z_q}{U_o} & \frac{-g(\sin\Theta_o)}{U_o} \\ M_u & M_\alpha & M_q & 0 \\ 0 & 0 & 1 & 0 \end{bmatrix} \begin{bmatrix} u \\ \alpha \\ q \\ \theta \end{bmatrix} + \begin{bmatrix} X_{\delta E} & X_{\delta T} \\ Z_{\delta E} & 0 \\ M_{\delta E} & 0 \\ 0 & 0 \end{bmatrix} \begin{bmatrix} \delta_E \\ \delta_T \end{bmatrix} \tag{3.57}
$$

The output variables are tentatively selected to be: (1) Linear airspeed (u), (2) Angle of attack (α), (3) Pitch rate (q), and (4) climb angle (γ). Recall—from definition—that, the climb angle is a function of pitch angle and angle of attack as:

$$
\gamma = \theta - \alpha \tag{3.58}
$$

This, the output equation is presented as:

$$
\begin{bmatrix} u \\ \alpha \\ q \\ \gamma \end{bmatrix} = \begin{bmatrix} 1 & 0 & 0 & 0 \\ 0 & 1 & 0 & 0 \\ 0 & 0 & 1 & 0 \\ 0 & -1 & 0 & 1 \end{bmatrix} \begin{bmatrix} u \\ \alpha \\ q \\ \theta \end{bmatrix} + \begin{bmatrix} 0 & 0 \\ 0 & 0 \\ 0 & 0 \\ 0 & 0 \end{bmatrix} \begin{bmatrix} \delta_E \\ \delta_T \end{bmatrix} \tag{3.59}
$$

In this longitudinal state space model, A is a 4×4 square matrix, the C is a 4×4 matrix, and B and D are 4×2 matrices, where in D, all elements are zero. The applications of this model will be presented in Sect. 3.4, as well as in Chap. 5 for longitudinal control.

3.2.4 Longitudinal Force and Moment Coefficients

The aerodynamic forces and moment coefficients can be modelled and linearized with respect to the stability and control derivatives. These coefficients are expressed in terms of flight parameters and control surface deflections. As explained in Chap. 1, since some flight variables such as airspeed and pitch rate have dimensions, they have been multiplied/divided by some variables to make the outcome non-dimensional.

$$
C_D = C_{D_o} + C_{D_\alpha}\alpha + C_{D_q}q\frac{C}{2U_1} + C_{D_{\dot\alpha}}\dot\alpha\frac{C}{2U_1} + C_{D_u}\frac{u}{U_1} + C_{D_{\delta_e}}\delta_e \tag{3.60}
$$

$$
C_L = C_{L_o} + C_{L_\alpha}\alpha + C_{L_q}q\frac{C}{2U_1} + C_{L_{\dot\alpha}}\dot\alpha\frac{C}{2U_1} + C_{L_u}\frac{u}{U_1} + C_{L_{\delta_e}}\delta_e \tag{3.61}
$$

$$
C_m = C_{m_o} + C_{m_\alpha}\alpha + C_{m_q}q\frac{C}{2U_1} + C_{m_{\dot\alpha}}\dot\alpha\frac{C}{2U_1} + C_{m_u}\frac{u}{U_1} + C_{m_{\delta_e}}\delta_e \tag{3.62}
$$

The significance and calculations of the longitudinal stability derivatives will be presented in the next few sections. Table 3.1 illustrates typical values of longitudinal non-dimensional stability and control derivatives for fixed-wing aircraft.

Table 3.1 Longitudinal non-dimensional stability and control derivatives

Derivative	Typical values (1/rad)		Prediction accuracy (%)
	Minimum	Maximum	
C_{D_u}	−0.01	+0.3	80
C_{L_u}	−0.2	+0.6	80
C_{m_u}	−0.4	+0.6	80
C_{D_α}	0	+2.0	90
C_{L_α}	+1.0	+8.0	95
C_{m_α}	0	−20	90
$C_{L_{\dot\alpha}}$	−5.0	+15.0	60
$C_{m_{\dot\alpha}}$	0	−20.0	60
C_{L_q}	0	+30.0	80
C_{m_q}	0	−90	80
$C_{L_{\delta_e}}$	0	+0.6	80
$C_{m_{\delta_e}}$	0	−4.0	80

Table 3.1 also demonstrates the prediction accuracy of the mathematical techniques presented to calculate the values of the longitudinal non-dimensional stability and control derivatives. As seen, the highest accuracy belongs to the technique to determine C_{L_α} (95%), but the lowest accuracy is for calculations of derivatives due to $C_{L_{\dot\alpha}}$ (about 60%).

Two better and more accurate techniques for calculations of stability and control derivatives are: (1) Wind tunnel tests, and (2) Flight tests. Using flight test results, one can identify aircraft dynamic parameters and characteristics [2] including stability derivatives. Various techniques have been developed, tested and illustrated [3, 4] for the determination of aircraft aerodynamic derivatives from flight test data. One popular approach is the maximum likelihood estimation scheme [5] which can accurately estimate the aircraft derivatives via flight tests.

3.3   Longitudinal Static Stability

3.3.1   Fundamentals

Longitudinal stability refers to the tendency of an aircraft to oppose any longitudinal disturbance, and to return to its equilibrium longitudinal state of motion, if disturbed in the xz plane. Longitudinal static stability is defined as to the tendency of an aircraft to develop forces and moments to oppose any longitudinal disturbance/perturbations. In a fixed-wing conventional air vehicle, the longitudinal stability is mainly provided by the horizontal tail.

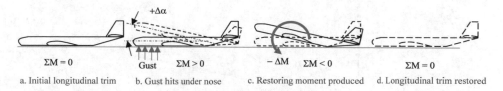

$\Sigma M = 0$ Gust $\Sigma M > 0$ $-\Delta M$ $\Sigma M < 0$ $\Sigma M = 0$

a. Initial longitudinal trim   b. Gust hits under nose   c. Restoring moment produced   d. Longitudinal trim restored

Fig. 3.2 A longitudinal disturbance and the aircraft reaction

The most important flight parameters in the longitudinal plane (xz) are: (1) Angle of attack (α), (2) Airspeed (V), (3) Altitude (h), (4) Pitch angle (θ), and (5) Climb angle (γ), and (6) Pitch rate (q). In a longitudinally statically stable aircraft, these parameters are expected to be maintained, if the aircraft is disturbed. All these parameters are inter-related, but they can be represented by the most sensitive one, which is the angle of attack. To have longitudinal static stability, aircraft needs to develop a restoring force and/or pitching moment, when it is displaced from its equilibrium point.

The primary factors that affect longitudinal stability can be realized by referring to Fig. 3.2 that shows the impact of a gust on an air vehicle and the reaction of a stable aircraft. Initially, the aircraft is at longitudinal trim where the sum of the moments about y-axis (cg) and the sum of the forces along x and z are zero. In Fig. 3.2a, the aircraft is cruising with a constant airspeed, constant altitude, and a constant angle of attack (α).

In Fig. 3.2b, a vertical gust hits under the fuselage nose and increases the angle of attack ($+\Delta\alpha$). After the gust disturbs the aircraft, sum of the moments about y-axis will no longer be zero. The reaction of a trimmed stable aircraft to this gust is illustrated in Fig. 3.2c, where a negative aerodynamic moment ($-\Delta M$) is automatically generated. This restoring moment is the opposition of aircraft to a positive angle of attack generated by gust. A restoring pitching moment is the aircraft reaction to oppose this longitudinal disturbance. After a few seconds, the aircraft finally returns back to initial longitudinal trim point (Fig. 3.2d).

3.3.2 Longitudinal Static Stability Derivatives

The pitching moment coefficient (C_m) is a good representative of the pitching moment. Consider the longitudinally statically stable aircraft shown in Fig. 3.3 with a negative slope of C_m versus angle attack. The aircraft is initially longitudinally trimmed at the trim point denoted by α_o that is $C_m = 0$. This is just one trim point. Suppose the aircraft nose suddenly encounter an upward gust such that the angle of attack is increased (i.e., $\alpha_o + \Delta\alpha$). This new angle of attack is corresponding to a new pitching moment coefficient ($-\Delta C_m$) which implies the aircraft develops a negative pitching moment ($-M$ or nose-down). This aircraft reaction will tend to rotate the aircraft back toward its initial trim angle of attack.

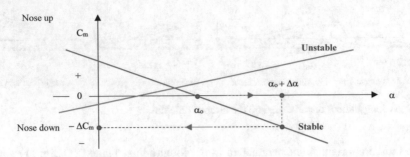

Fig. 3.3 Pitching moment curves for statically longitudinally stable and unstable aircraft

For the same disturbance, a statically longitudinally unstable aircraft would develop a positive pitching moment, since the C_m–α curve is positive. This aircraft natural reaction will tend to rotate the aircraft further away from the initial trim angle of attack. If the pitching moment causes the nose to rise further after a nose-up disturbance, then the aircraft will continue to pitch up and is statically longitudinally unstable.

Thus, when the aircraft angle of attack is increased by a disturbance, a negative (i.e., nose down) pitching moment is desired to oppose the disturbance and restore the trim point. In addition, when the aircraft angle of attack is decreased, a positive (i.e., nose up) pitching moment is desired. The longitudinal static stability requirement is formulated as the rate of change of pitching moment coefficient (C_m) with respect to the angle of attack (α) or $C_{m\alpha}$. Thus, the requirement for longitudinal static stability is to have a negative value for this parameter.

$$C_{m_\alpha} = \frac{dC_m}{d\alpha} < 0 \tag{3.63}$$

The derivative C_{m_α} is called the static longitudinal stability derivative. The higher the negative value, the aircraft will be more longitudinally statically stable. Typical value for $C_{m\alpha}$ of a longitudinally statically stable aircraft can range from −1 to −3 1/rad. There are more parameters like this stability derivative that influence the longitudinal static stability characteristics of the aircraft. They will be presented in the next section.

The most important derivative in longitudinal static stability requirements is C_{m_α}. The pitching moment coefficient is a function of the lift coefficient, and this also can be used to evaluate the aircraft static stability. Since, the angle of attack is directly a function of lift coefficient, we can see the same feature (i.e., negative slope) in analyzing longitudinal static stability. Mathematically, a stable system must have a pitching moment versus lift coefficient curve with a negative slope.

For an aircraft with a fixed horizontal tail, the aircraft static longitudinal stability derivative is determined [6] by adding wing, fuselage, and horizontal tail contributions as:

$$C_{m_\alpha} = C_{L_{\alpha_{wf}}} \left(\frac{x_{cg}}{\overline{C}} - \frac{x_{ac_{wf}}}{\overline{C}} \right) - C_{L_{\alpha_h}} \eta_h \frac{S_h}{S} \left(\frac{l_h}{\overline{C}} \right) \left(1 - \frac{d\varepsilon}{d\alpha} \right) \qquad (3.64)$$

where $C_{L_{\alpha_{wf}}}$ and $C_{L_{\alpha_h}}$ denote wing-fuselage and horizontal tail lift curve slope respectively, $x_{ac_{wf}}$, x_{ac_h}, and x_{cg} refer to distance between wing-fuselage aerodynamic center, tail aerodynamic center, and aircraft center of gravity to a reference line (e.g., wing leading edge at the mean aerodynamic chord location) respectively. The parameters $\frac{d\varepsilon}{d\alpha}$ is the downwash slope at the tail, and η_h is the horizontal tail dynamic pressure ratio.

Equation (3.64) applies to a fixed wing aircraft with both aft-tail or canard. For an aircraft with canard (i.e., foreplane), l_h is the distance between canard ac to the aircraft cg (i.e., canard pitching moment arm).

Figure 3.4 illustrates the wing-fuselage aerodynamic center, horizontal tail aerodynamic center, and the aircraft center of gravity (cg). The relation between these three centers plus the tail surface area have the direct impact on this derivative (C_{m_α}). The reference line for these three centers is the wing leading edge at the mean aerodynamic chord location.

From Eq. (3.64), one can conclude that, in order to make an aircraft longitudinally statically more stable, there are three primary solutions: (1) Increase the size of horizontal tail larger, (2) Move the horizontal tail further away from aircraft cg, (3) Make the aircraft cg further away aft of wing-fuselage aerodynamic center. A large tail located a long distance aft of the aircraft cg is a powerful longitudinal stabilizer. More discussions on the third solution will be presented in the following section.

The static longitudinal stability of the aircraft is also influenced by the propulsion system. However, the contribution of the engine (jet or prop-driven) to the static longitudinal stability is not as significant as the contributions of wing and tail, and is difficult to estimate. Hence, we neglected its contribution; Ref. [7] provides techniques to estimate the engine and propeller contribution. The degree of static stability can be controlled by proper selection of the horizontal tail surface area and its arm to aircraft cg.

In Eq. (3.64), distances are non-dimensionalized with respect to mean aerodynamic chord $(\overline{C})$ to make the value more convenient to compare. The mean aerodynamic chord

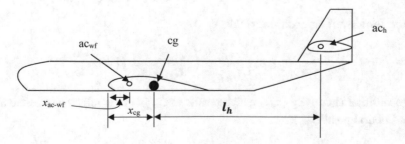

Fig. 3.4 Wing and tail aerodynamic centers and their distances to a reference line

is utilized in the calculation of aspect ratio:

$$AR = \frac{b}{\overline{C}} \tag{3.65}$$

Moreover, the wing planform area can be determined by multiplying wing span (b) and mean aerodynamic chord:

$$S = b\overline{C} \tag{3.66}$$

The mean aerodynamic chord ($\overline{C}$) is a function of wing tip chord, wing root chord and wing span. A wing at every section has its own aerodynamic center. In Fig. 3.4, the wing aerodynamic center is located at the mean aerodynamic chord location (not root chord or tip chord).

3.3.3   Neutral Point

This section is devoted to another indicator of longitudinal static stability, the aircraft neutral point. The definition of this unique point and technique to determine its location is presented. The expression for $C_{m\alpha}$ (Eq. 3.64) indicates that its sign and value depend upon the center of gravity position as well as the tail characteristics. From the values and minus sign of the last term, it is clear that the tail has a powerful stabilizing contribution. However, the center of gravity position may have a stabilizing or destabilizing contribution, due to its relationship with respect to wing-fuselage aerodynamic center $x_{ac_{wf}}$.

If C_{m_α} is negative, the aircraft is longitudinally statically stable; but if C_{m_α} is positive, the aircraft is longitudinally statically unstable. However, if C_{m_α} is zero, the aircraft is assumed to be longitudinally statically neutrally stable. We are interested to find out at what cg location, the value of C_{m_α} becomes zero. When setting $C_{m_\alpha} = 0$ in Eq. (3.64), we obtain:

$$0 = C_{L_{\alpha_{wf}}} \left(\frac{x_{cg}}{\overline{C}} - \frac{x_{ac_{wf}}}{\overline{C}} \right) - C_{L_{\alpha_h}} \eta_h \frac{S_h}{S} \left(\frac{x_{ac_h}}{\overline{C}} - \frac{x_{cg}}{\overline{C}} \right) \left(1 - \frac{d\varepsilon}{d\alpha} \right) \tag{3.67}$$

Solving for aircraft cg position, yields,

$$\frac{x_{cg}}{\overline{C}} = \frac{x_{ac_{wf}}}{\overline{C}} + \eta_h \frac{C_{L_{\alpha_h}}}{C_{L_{\alpha_{wf}}}} \frac{S_h}{S} \left(\frac{x_{ac_h}}{\overline{C}} - \frac{x_{cg}}{\overline{C}} \right) \left(1 - \frac{d\varepsilon}{d\alpha} \right) \tag{3.68}$$

The location of aircraft cg (x_{cg}) which results in $C_{m_\alpha} = 0$, is referred to as the aircraft *stick-fixed* neural point (x_{np}).

$$\frac{x_{np}}{\overline{C}} = \frac{x_{ac_{wf}}}{\overline{C}} + \eta_h \frac{C_{L_{\alpha_h}}}{C_{L_{\alpha_{wf}}}} V_H \left(1 - \frac{d\varepsilon}{d\alpha} \right) \tag{3.69}$$

where V_H is the horizontal tail volume coefficient. The term "*stick-fixed*" is utilized here to indicate that the stick (that controls the elevator) should be kept by pilot hand. If an aircraft uses an elevator tab such that stick can be left to stay free, Eq. (3.69) should be modified (mainly $C_{L_{\alpha_h}}$ and x_{ac_h}). When stick is free to move, the elevator is free to move too. However, the elevator tab keeps the elevator in place, by trimming the pitching moment about elevator hinge. To allow the stick to be free, pilot first needs to select the location of the elevator by hand. Then, leave it to the tab to maintain the elevator deflection. It is reminded that Eqs. (3.64) and (3.69) do not include the engine thrust effect and propeller effect.

The aircraft cg varies during the course of its flight operation; thus, it is important to know, what the most *aft limit* is before aircraft becomes longitudinally statically unstable. Movement of the aircraft cg aft of the neutral point causes the aircraft to be statically longitudinally unstable. The location of aircraft np can be calculated from Eq. (3.69), or can be determined with flight tests [6] measurement.

To experimentally determine np, one needs to find the cg at which zero δ_E is needed [6], to change the angle of attack (i.e., $d\delta_E/dC_L = 0$). Since it is unsafe to fly at that cg, we must determine the $d\delta_E/dC_L$ at each of various aft cg's (between forward and aft limits) and extrapolate to zero. To do that, measure and plot δ_E versus C_L at the different cg's and then, measure the slope. Each slope provides one in Fig. 3.5. Recall that, different values of C_L mean different (constant) speeds, at each cg flown. Typical values for the aircraft np in subsonic speeds lie in the range of 40–60% MAC. However, for supersonic speeds, the np moves to 60–70% MAC. The reason is that, the wing ac moves from about 25% MAC at subsonic speeds to about 50% MAC at supersonic speeds. Thus, the aircraft np moves back, after airspeed exceeds the speed of sound.

In designing an aircraft, and during aircraft loading, placing the aircraft cg further away forward of aircraft np makes the aircraft more longitudinally stable. Typical positions for the wing-fuselage aerodynamic center, aircraft center of gravity, and neutral point of a longitudinally stable fixed-wing aircraft are demonstrated in Fig. 3.6.

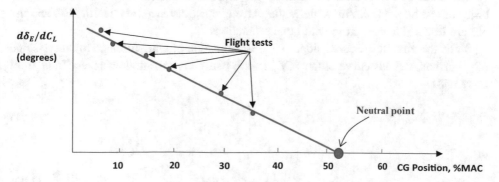

Fig. 3.5 Variations of $d\delta_E/dC_L$ w.r.t. aircraft CG position

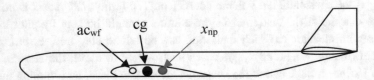

Fig. 3.6 Wing-fuselage aerodynamic center, aircraft center of gravity, and neutral point

Since, when the aircraft cg is located at the np, the C_{m_α} will be zero, some textbooks refer to the np as the aircraft aerodynamic center. This is due to the fact that, in an airfoil and a wing, variations of C_m about ac with respect to angle of attack (α) is zero. This similarity allows some references to employ a similar title for aircraft np.

The static longitudinal stability is examined through the sign of the derivative Cm_α or the location of the aircraft neutral point. When the Cm_α is negative, or when the np is behind the aircraft cg, the aircraft is said to be statically longitudinally stable.

The distance between the aircraft cg to the aircraft neutral point (np) is a margin of flight safety. This distance, when non-dimensionalized with respect to mean aerodynamic chord, is referred to as the *static margin*:

$$SM = \frac{x_{np}}{\overline{C}} - \frac{x_{cg}}{\overline{C}} \tag{3.70}$$

The np is the most aft allowable cg location (to satisfy longitudinal static stability requirement). For a civil aircraft, it is recommended to have a positive static margin of at least 10% of the mean aerodynamic chord. On the Boeing 777-200, the cg positions from 14 to 44% of MAC are allowed. Thus, if a minimum SM of 10% is assumed, the np should be at least behind 54% of MAC.

However, for a fighter aircraft, it is recommended to have the aircraft cg behind the aircraft np (or NP); which means a negative static margin. Although, this adoption makes the fighter longitudinally unstable, but it will be highly maneuverable.

The distance between the NP and CG has a profound impact on the longitudinal stability of the air vehicle. Air vehicles that have a small distance between the CG and the NP are less stable than those with large separations.

Since the aircraft lift curve slope (C_{L_α}) is a function of wing-fuselage lift curve slope ($C_{L_{\alpha_{wf}}}$) and tail lift curve slope ($C_{L_{\alpha_h}}$), one can derive the following equation from Eq. (3.58).

$$C_{m_\alpha} = C_{L_\alpha} \left[\frac{x_{cg}}{\overline{C}} - \frac{x_{np}}{\overline{C}} \right] \tag{3.71}$$

or

$$C_{m_\alpha} = -C_{L_\alpha}[SM] \tag{3.72}$$

This equation indicates the direct and linear relation between static margin and the static longitudinal stability derivative. Features of stick free and stick fixed control modes are discussed in Chap. 5.

Example 3.1 Consider a fixed-wing General Aviation aircraft—shown in Fig. 3.7—with a maximum takeoff weight of 2600 lb, a wing planform area of 175 ft^2, and the following characteristics

$$\frac{d\varepsilon}{d\alpha} = 0.4; \ C_{L_{\alpha_{wf}}} = 5.4 \ \frac{1}{\text{rad}}; \ C_{L_{\alpha_h}} = 4.6 \ \frac{1}{\text{rad}}; \ b = 36 \ \text{ft}; \ \eta_h = 0.97; \ S_h = 44 \ \text{ft}^2$$

Determine longitudinal static stability derivative; C_{m_α}, neutral point position (np), and static margin (SM). Is this aircraft longitudinally statically stable?

Solution

$$S = b\overline{C} \Rightarrow \overline{C} = \frac{S}{b} = \frac{175}{36} = 4.86 \ \text{ft} \tag{3.66}$$

From figure, the tail ac is 18 ft from wing leading edge ($x_{ac_h} = 18$ ft), the wing-fuselage ac is 1.2 ft from wing leading edge, while aircraft cg is 1.8 ft from wing leading edge. The longitudinal static stability derivative is:

$$C_{m_\alpha} = C_{L_{\alpha_{wf}}}\left(\frac{x_{cg}}{\overline{C}} - \frac{x_{ac_{wf}}}{\overline{C}}\right) - C_{L_{\alpha_h}}\eta_h\frac{S_h}{S}\left(\frac{x_{ac_h}}{\overline{C}} - \frac{x_{cg}}{\overline{C}}\right)\left(1 - \frac{d\varepsilon}{d\alpha}\right) \tag{3.64}$$

$$C_{m_\alpha} = 5.4\left(\frac{1.8}{4.86} - \frac{1.2}{4.86}\right) - 4.6 \times 0.97 \times \frac{44}{175}\left(\frac{18}{4.86} - \frac{1.8}{4.86}\right)(1 - 0.4) = -1.58 \ \frac{1}{\text{rad}}$$

The horizontal tail volume ratio:

$$\overline{V}_h = \frac{l_h S_h}{\overline{C}S} = \frac{18 \times 44}{4.86 \times 175} = 0.838 \tag{2.13}$$

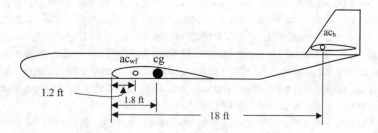

Fig. 3.7 General aviation aircraft for Example 3.1 (not to scale)

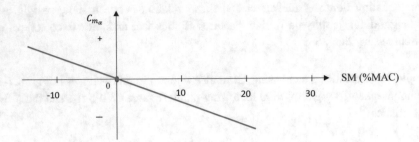

Fig. 3.8 Relationship between static margin and C_{m_α}

The neutral point:

$$\frac{x_{np}}{\bar{C}} = \frac{x_{ac_{wf}}}{\bar{C}} + \eta_h \frac{C_{L_{\alpha_h}}}{C_{L_{\alpha_{wf}}}} V_h \left(1 - \frac{d\varepsilon}{d\alpha}\right)$$

$$= \frac{1.2}{4.86} + 0.97 \times \frac{4.6}{5.4} \times 0.838 \times (1 - 0.4) = 0.662 \qquad (3.69)$$

The neutral point is at 66.2% MAC. Since the aircraft cg is at 30% MAC ($\frac{x_{cg}}{\bar{C}} = 0.3$), the neutral point is behind center of gravity. The static margin:

$$SM = \frac{x_{np}}{\bar{C}} - \frac{x_{cg}}{\bar{C}} = 0.662 - \frac{1.8}{4.8} = 0.215 = 29.2\% \qquad (3.70)$$

The aircraft is longitudinally statically stable, since the C_{m_α} is negative. We can derive the same conclusion, since, np is behind cg, with a static margin of 29.2% MAC.

Relation between two indicators of static longitudinal stability (1. Static margin, 2. C_{m_α}) are illustrated in Fig. 3.8. As the static margin is increased (more positive), C_{m_α} becomes more negative, so the aircraft will become statically longitudinally more stable.

Theoretical typical location for the wing aerodynamic center (x_{ac}) is about quarter chord (1/4 or 25% C). Hence, to have at least 5% static margin, the most aft cg of the aircraft should be located aft of 30% MAC. Table 3.2 shows the most aft cg for several aircraft in terms of percent MAC. As you see, for most subsonic aircraft, the most aft cg is located between 30 and 40% MAC. However, for the case of Concorde, the supersonic transport aircraft, the most aft cg is located at 59% MAC. This is due to the fact that, at supersonic airspeeds (M = 2.2), the wing aerodynamic center will be about 50% MAC due to oblique shock waves. Hence, the cg should be aft of this point (e.g., about 65%), and neutral point is aft of about 70% MAC.

From this relation, for any aircraft, the center of gravity (cg) must be ahead of the aircraft neutral point for the pitching moment-lift curve to remain negative. The horizontal tail is an important lifting surface for stability of an air vehicle. Adding a second

Table 3.2 The most aft cg location for several aircraft

No	Aircraft	Engine(s)	m_{TO} (kg)	Most aft cg (% MAC)
1	Cessna 172	Single piston	1111	36.5
2	Cessna Citation III	Twin turbofan	9527	31
3	Beechcraft B-45	Single turboprop	1950	28
4	Pilatus PC-12	Single turboprop	4740	46
5	Piper PA-30 Comanche	Twin piston	1690	27.8
6	Gulfstream G550	Twin turbofan	41,277	45
7	Lockheed C-130E	Four turboprop	70,300	30
8	DC-9-10	Twin turbofan	41,100	40
9	Boeing 737-100	Twin turbofan	50,300	31
10	Boeing 747-200	Four turbofan	377,842	32
11	Boeing 777-200	Twin turbofan	247,200	44
12	Lockheed C-141	Four turbofan	147,000	32
13	Lockheed C-5A	Four turbofan	381,000	41
14	Concorde	Four turbojet	185,700	59

lifting surface (first is the wing) away from the cg; either behind (aft horizontal tail), or ahead (canard) has the effect of moving the overall aircraft neutral point rearward, which increases stability. For this reason, the horizontal tail or canard may be regarded as the horizontal stabilizer.

3.4 Longitudinal Dynamic Stability

3.4.1 Fundamentals

Aircraft dynamic stability is defined as the tendency of the aircraft to return to the equilibrium state of motion, if the flight is disturbed. We are mainly concerned with the time history of the motion of the aircraft after it is disturbed from its equilibrium point. As we discussed earlier, the aircraft static stability is the opposition of the aircraft to any flight disturbance. The static stability is a pre-requisite to the dynamic stability; the aircraft dynamic stability requires the aircraft to be statically stable. If an aircraft is statically stable, it may or may not be dynamically stable. However, if an aircraft is dynamically stable, it has to be statically stable too.

Longitudinal dynamic stability is defined as the tendency of an aircraft to oppose any longitudinal disturbance (e.g., up/down gust), and to return to its equilibrium longitudinal state of motion, if disturbed. Because of the damping forces and natural friction in real systems, longitudinal dynamic stability is usually, but not always, present if the system is longitudinal statically stable. In this section, we shall examine the longitudinal motion of an aircraft, after it is disturbed from its trim/equilibrium state.

The behavior of an aircraft with regards to a disturbance to its angle of attack, which is statically stable, and also dynamically stable is exhibited in Fig. 3.9a. In this aircraft, the trim angle of attack is initially zero. A gust under fuselage nose increases the angle of attack to about 1.4°. Since the aircraft is statically stable, it opposes this disturbance, and tries to return the angle of attack to trim value. However, since the aircraft is dynamically stable too, it will finally succeed to return the angle of attack to zero.

The behavior of an aircraft with regards to a disturbance to its angle of attack, which is statically stable, but dynamically unstable is exhibited in Fig. 3.9b. In this aircraft, the trim angle of attack is initially zero. A gust under fuselage nose increases the angle of attack to about 2°. Since the aircraft is statically stable, it opposes this disturbance, and tries to return the angle of attack to trim value. However, since the aircraft is dynamically unstable, it will not succeed to return the angle of attack to zero and diverges. Every time, the aircraft gets further away from its trim value (divergence), until the aircraft crashes.

A comparison between the reaction of two dynamically stable aircraft with different levels of stability is presented in Fig. 3.10. In both aircraft, the trim angle of attack is initially zero, and a gust hits under fuselage nose and suddenly increases the angle of attack. This aircraft in Fig. 3.10a is much more stable than the aircraft in Fig. 3.10b. The aircraft in Fig. 3.10a experiences a more powerful gust that increases its angle of attack to about 3°. Since the aircraft is highly dynamically statically stable, it returns to its trim zero angle of attack in only about 8 s.

However, the aircraft in Fig. 3.10b experiences a weaker gust that increases its angle of attack to about 1°. But it takes a much longer time (about 180 s) to damp the disturbance and to return to its trim zero angle of attack.

Dynamic stability can be augmented by aeromechanical systems, such as an autopilot which primarily is meant to control the aircraft. A popular stability augmenter for large transport aircraft is the pitch damper which augments the longitudinal dynamic stability. However, dynamic instability may arise, if the controller and feedback in the control

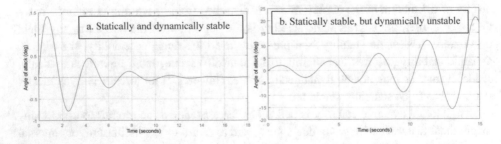

Fig. 3.9 Two statically stable aircraft, one dynamically stable, but one dynamically unstable

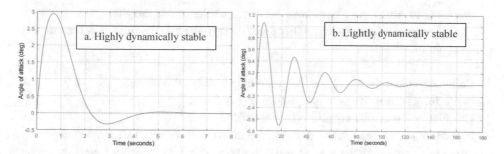

Fig. 3.10 Two dynamically stable aircraft with different levels of stability

system is improperly designed or compensated. This can lead to an amplification of the instability instead of damping.

3.4.2 Longitudinal Dynamic Stability Analysis Technique

Analysis of longitudinal dynamic stability is much more complex than that for the longitudinal static stability and requires a lot of calculations. The analysis may be conducted in either of two domains: (1) Frequency (s) domain, (2) Time (t) domain.

In frequency domain analysis, transfer functions (in s-domain) are utilized. In practice, only denominator of transfer functions suffices. As derived in Sect. 3.3, the denominators of all longitudinal transfer functions are similar, and is a polynomial as a function of s. When this fourth-order equation is equated to zero, we obtain the following form:

$$A_1 s^4 + B_1 s^3 + C_1 s^2 + D_1 s + E_1 = 0 \tag{3.73}$$

where the coefficients A_1, B_1, C_1, D_1, and E_1 are real quantities and have been already presented. Since the longitudinal absolute stability can be analyzed directly from these coefficients, this equation is referred to as the longitudinal characteristic equation. The order, coefficients, and roots of this equation reveal a lot of information about characteristics of the aircraft with regards to longitudinal dynamic stability.

In time domain analysis, either the governing differential equations of motion, or state space model of dynamics are employed. In practice, only matrix A (i.e., state matrix) for longitudinal motion (A_{lon}) suffices:

$$A_{lon} = \begin{bmatrix} X_u & X_\alpha & 0 & -g\cos\Theta_o \\ \frac{Z_u}{U_o} & \frac{Z_\alpha}{U_o} & 1 + \frac{Z_q}{U_o} & \frac{-g(\sin\Theta_o)}{U_o} \\ M_u & M_\alpha & M_q & 0 \\ 0 & 0 & 1 & 0 \end{bmatrix} \tag{3.74}$$

Elements of matrix A_{lon} have been presented earlier in this chapter.

In the longitudinal dynamic stability analysis in s-domain, the following four steps are conducted:

1. Calculate values for all aircraft longitudinal stability derivatives;
2. Write longitudinal characteristic equation (Eq. 3.73);
3. Calculate roots of this equation (four roots);
4. If roots have all negative real parts, the air vehicle is said to be dynamically longitudinally stable. If there is at least one root with a positive real part, the aircraft is dynamically longitudinally unstable. However, if at least real part of one root is positive, and all other roots have negative real parts, the aircraft is dynamically longitudinally neutrally stable.

In the longitudinal dynamic stability analysis in time-domain, the following four steps are conducted:

1. Calculate values for all aircraft longitudinal stability derivatives;
2. Write state matrix (i.e., A) of longitudinal state space model (Eq. 3.74);
3. Calculate eigenvalues of this matrix (four eigenvalues);
4. If eigenvalues have all negative real parts, the air vehicle is said to be dynamically longitudinally stable. If there is at least one eigenvalue with a positive real part, the aircraft is dynamically longitudinally unstable. However, if at least real part of one eigenvalue is positive, and all other eigenvalues have negative real parts, the aircraft is dynamically longitudinally neutrally stable.

Example 3.2 A GA aircraft has the following angle-of-attack-to-elevator-deflection transfer function for a cruising flight:

$$\frac{\alpha(s)}{\delta_E(s)} = -\frac{35s^3 + 7020s^2 + 340s + 280}{205s^4 + 2030s^3 + 5800s^2 + 280s + 164}$$

Is the aircraft dynamically longitudinally stable? Why?

Solution

The longitudinal characteristic equation (i.e., denominator) is:

$$205s^4 + 2030s^3 + 5800s^2 + 280s + 164 = 0$$

Four roots of longitudinal characteristic equation are: $-4.93 \pm 1.87i$; $-0.0194 \pm 0.168i$. All roots are complex (two complex pairs). Since all roots have negative real parts, the aircraft is dynamically longitudinally stable.

Example 3.3 A GA aircraft has the following longitudinal state space model for a cruising flight at 200 ft/s:

$$
\begin{bmatrix} \dot{u} \\ \dot{\alpha} \\ \dot{q} \\ \dot{\theta} \end{bmatrix} = \begin{bmatrix} -0.04 & 19 & 0 & -32.2 \\ -0.05 & -1.6 & 0.94 & 0 \\ 0 & -2.2 & -2.4 & 0 \\ 0 & 0 & 1 & 0 \end{bmatrix} \begin{bmatrix} u \\ \alpha \\ q \\ \theta \end{bmatrix} + \begin{bmatrix} 0 & 12 \\ -18 & 0 \\ -5.4 & 0 \\ 0 & 0 \end{bmatrix} \begin{bmatrix} \delta_E \\ \delta_T \end{bmatrix}
$$

$$
\begin{bmatrix} u \\ \alpha \\ q \\ \gamma \end{bmatrix} = \begin{bmatrix} 1 & 0 & 0 & 0 \\ 0 & 1 & 0 & 0 \\ 0 & 0 & 1 & 0 \\ 0 & -1 & 0 & 1 \end{bmatrix} \begin{bmatrix} u \\ \alpha \\ q \\ \theta \end{bmatrix} + \begin{bmatrix} 0 & 0 \\ 0 & 0 \\ 0 & 0 \\ 0 & 0 \end{bmatrix} \begin{bmatrix} \delta_E \\ \delta_T \end{bmatrix}
$$

Is the aircraft dynamically longitudinally stable? Why?

Solution

The state matrix (A) of the longitudinal state space model is:

$$
\begin{bmatrix} -0.04 & 19 & 0 & -32.2 \\ -0.05 & -1.6 & 0.94 & 0 \\ 0 & -2.2 & -2.4 & 0 \\ 0 & 0 & 1 & 0 \end{bmatrix} \tag{3.74}
$$

Four eigenvalues of this matrix are: $-1.99 \pm 1.51i$; $-0.021 \pm 0.75i$. All eigenvalues are complex (two complex pairs). Since all eigenvalues have negative real parts, the aircraft is dynamically longitudinally stable.

3.4.3 Longitudinal Response Modes

When a dynamically longitudinally stable aircraft responds to a longitudinal disturbance, longitudinal flight variables (e.g., airspeed, angle of attack, and altitude) will often follow a damped oscillatory (often a sinusoidally) motion. However, variations of some flight parameters have a much higher frequency, while variations of others have a much lower frequency. Thus, in practice, there are often two distinct modes of responses: (1) Short-period mode, and (2) Long-period or Phugoid mode. These two modes are distinct, but they happen simultaneously.

In general, the short period mode is much better damped than the phugoid mode. The short period mode is often damped much faster (about a few seconds) than the phugoid mode (about a few minutes). The mode that is lightly damped has a long period, and the mode that is highly damped has a short period. The short period mode natural frequency is more than an order of magnitude larger than that of the phugoid mode.

In majority of fixed-wing aircraft configurations, variations of angle of attack, pitch angle, and pitch rate have a short period mode of motion, while that for airspeed and altitude have a long period mode of motion (Fig. 3.11).

As seen in Fig. 3.11a, a *short period* mode has a scale of a few seconds. The aircraft experiences a perturbation (impulse) in angle of attack of $+2°$, after a gust hit under the fuselage nose. This perturbation is damped out in about 3 s, and the aircraft returns back to the steady-state value of angle of attack (here, zero). During this mode of reaction, the airspeed remains nearly constant, while angle of attack and pitch angle vary. This damped oscillatory motion has a period of 1.31 s (i.e., short).

As seen in Fig. 3.11b, a *phugoid mode* has a scale of a few minutes. The aircraft experiences a perturbation (impulse) in airspeed of $+3$ knot, after a gust hit back of the fuselage. This perturbation is damped out in about 80 s, and the aircraft returns back to the steady-state value of airspeed. During this mode of reaction, the angle of attack remains constant, while airspeed and altitude vary. This damped oscillatory motion has a period of 5.1 s (i.e., long).

Any higher order transfer function may be broken down to the combination of a number of first order and second transfer functions. For instance, any proper fourth order transfer function (here, longitudinal transfer functions) can be reformatted as the combination of two second order transfer functions:

$$\frac{Out(s)}{In(s)} = \frac{Num(s)}{A_1 s^4 + B_1 s^3 + C_1 s^2 + D_1 s + E_1}$$
$$= \frac{Num_1(s)}{s^2 + a_1 s + b_1} + \frac{Num_2(s)}{s^2 + a_2 s + b_2} \qquad (3.75)$$

where Num stands for numerator. In this format, one of the second order transfer functions represents the short-period mode, and the other one represents the long period or phugoid mode.

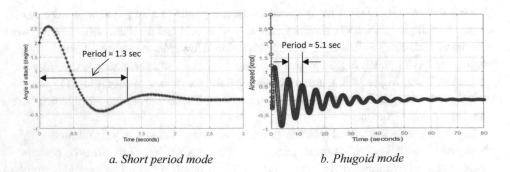

a. Short period mode   b. Phugoid mode

Fig. 3.11 Typical short period and phugoid modes

Equations (3.5)–(3.10) can be applied to both longitudinal modes (i.e., short-period and phugoid). However, before application, make sure that: (1) The aircraft is longitudinally stable, (2) All roots are complex conjugates. If any of these two conditions are not met, these equations cannot be applied. In another word, a dynamically longitudinally unstable aircraft, we are interested in the time it takes for the initial amplitude or disturbance to double. Table 3.3 provides longitudinal *handling qualities* [7] in terms of specified values of damping ratio of short period mode at three levels and three flight phases.

According to flying quality requirements [7], the values of damping ratio of phugoid mode shown in Table 3.4 must be achieved.

Example 3.4 A large transport aircraft has the following speed-to-elevator-deflection transfer function for a cruising flight:

$$\frac{u(s)}{\delta_E(s)} = \frac{-24s^2 + 3140s + 13{,}800}{805s^4 + 920s^3 + 1580s^2 + 165s + 150}$$

If the aircraft is dynamically longitudinally stable, determine features of phugoid and short period modes.

Solution
The longitudinal characteristic equation (i.e., denominator) is:

$$805s^4 + 920s^3 + 1580s^2 + 165s + 150 = 0$$

Table 3.3 Short period mode damping ratio specification

Flight phase	Short period damping ratio (ζ_s)					
	Level 1		Level 2		Level 3	
	Minimum	Maximum	Minimum	Maximum	Minimum	Maximum
A	0.35	1.3	0.25	2.0	0.15	No maximum
B	0.3	2.0	0.2	2.0	0.15	No maximum
C	0.35	1.3	0.25	2.0	0.15	No maximum

Table 3.4 Phugoid mode requirement

Level of acceptability	Requirement
1	Damping ratio of phugoid mode (ζ_{ph}) ≥ 0.04
2	Damping ratio of phugoid mode (ζ_{ph}) ≥ 0.0
3	The time-to-double the amplitude of at least 55 s

Four roots of longitudinal characteristic equation are: $-0.546 \pm 1.23i$; $-0.025 \pm 0.32i$. All four roots are imaginary (i.e., two complex pairs). Since all roots have negative real parts, the aircraft is dynamically longitudinally stable.

The roots of the **first mode** are:

$$s_{1,2} = \eta \pm i\omega = -0.546 \pm 1.23i \qquad (3.6)$$

Since real part (η), and imaginary part (ω) are:

$$s_{1,2} = -\zeta\omega_n \pm i\omega_n\sqrt{1-\zeta^2} \qquad (3.5)$$

We can equate real parts and imaginary parts as:

$$\eta = -0.546 = -\zeta\omega_n \qquad (3.7)$$

$$\omega = 1.23 = \omega_n\sqrt{1-\zeta^2} \qquad (3.8)$$

These two equations are solved simultaneously to determine natural frequency (ω_n) and damping ratio (ζ):

$$\zeta = 0.406; \omega_n = 1.35 \frac{\text{rad}}{\text{s}}$$

Period of this mode:

$$T = \frac{2\pi}{\omega} = \frac{2\pi}{1.23} = 5.11\,\text{s} \qquad (3.9)$$

The roots of the **second mode** are:

$$s_{1,2} = \eta \pm i\omega = -0.025 \pm 0.32i \qquad (3.6)$$

Since real part (η), and imaginary part (ω) are:

$$s_{1,2} = -\zeta\omega_n \pm i\omega_n\sqrt{1-\zeta^2} \qquad (3.5)$$

We can equate real parts and imaginary parts as:

$$\eta = -0.025 = -\zeta\omega_n \qquad (3.7)$$

$$\omega = 0.32 = \omega_n\sqrt{1-\zeta^2} \qquad (3.8)$$

These two equations are solved simultaneously to determine natural frequency (ω_n) and damping ratio (ζ):

$$\zeta = 0.078; \omega_n = 0.321 \frac{rad}{s}$$

Period of this mode:

$$T = \frac{2\pi}{\omega} = \frac{2\pi}{0.32} = 19.6\,s \qquad (3.9)$$

The first mode has a period of 5.1 s (shorter), while second mode has a period of 19.6 s (longer). Hence, the first mode represents the short period mode, while the second mode represent the phugoid mode. In summary,

The **short period** mode:

$$\zeta_{sp} = 0.406; \omega_{n_{sp}} = 1.35 \frac{rad}{s}; T_{sp} = 5.11\,s$$

The **phugoid period** mode:

$$\zeta_{ph} = 0.078; \omega_{n_{ph}} = 0.321 \frac{rad}{s}; T_{ph} = 19.6\,s$$

3.4.4 Longitudinal Modes Approximation

The characteristics of short period and phugoid modes for longitudinally stable aircraft can be predicted from an approximation to the longitudinal equations of motion. In this approximation, natural frequency and damping ratio are expressed as functions of dimensional stability derivatives.

An approximation to the phugoid mode is obtained by assuming that the change in angle of attack is zero, and by neglecting the pitching moment equation. Thus, for the *phugoid* mode, one can obtain:

$$\omega_{n_{ph}} = \sqrt{\frac{-Z_u g}{U_o}} \qquad (3.76)$$

$$\zeta_{ph} = \frac{-X_u}{2\omega_{n_{ph}}} \qquad (3.77)$$

An approximation to the short-period mode is obtained by dropping the X-force equation and assuming the change in airspeed is zero. Thus, for the *short period* mode, one can obtain:

$$\omega_{n_{sp}} = \sqrt{\frac{Z_\alpha M_q}{U_o} - M_\alpha} \qquad (3.78)$$

$$\zeta_{sp} = \frac{-1}{2\omega_{n_{sp}}}\left[M_q + M_{\dot{\alpha}} + \frac{Z_\alpha}{U_o}\right]$$ (3.79)

These equations provide a reasonable and good approximation (about 90% accuracy) for subsonic speeds with low angles of attack.

3.5   Aircraft Reaction to Longitudinal Disturbances

In order for an aircraft to feature dynamic stability, restoring forces/moments must be developed such that the flight motion disturbance is damped. Dynamic stability is created by aerodynamic forces and moments that are proportional to the rate and reaction of the various surfaces such as wing and tail. Each aircraft configuration has a unique contribution to the overall response to any disturbance. In this section, the aircraft responses to various longitudinal disturbance are discussed.

The perturbed-state forces and moments—created by disturbance—are temporary and additional to steady-state values of forces and moments. The perturbed-state forces and moments are temporarily generated to react to a disturbance, while the steady-state forces and moments are essential to trim the aircraft. When the aircraft returns to longitudinal trim condition, the perturbed-state forces and moments disappear.

Dynamic stability can be translated into aircraft disturbance damping power, which is mathematically modelled as the stability derivatives. There are mainly four direct longitudinal disturbance: (1) Disturbance to angle of attack ($\Delta\alpha$), (2) Disturbance to forward airspeed (Δu), (3) Disturbance to pitch angle ($\Delta\theta$), and pitch rate (Δq), (4) Disturbance to altitude (Δh or $\dot{\alpha}$).

Primary longitudinal stability derivatives are C_{m_q}; C_{L_q}; C_{D_q}; C_{m_u}; $C_{m_{\dot{\alpha}}}$; $C_{L_{\dot{\alpha}}}$; $C_{D_{\dot{\alpha}}}$; C_{L_u}; and C_{D_u}, which less or more represent aircraft reaction to longitudinal disturbances. The importance and application of these derivatives will be addressed in the following section.

3.5.1   Aircraft Reaction to Disturbance in Angle of Attack

The disturbance in angle of attack will cause three separate and parallel aircraft reactions: (1) Aerodynamic Pitching moment, (2) Lift, (3) Drag. These perturbed-state moment and forces are discussed here. An angle of attack perturbation will induce local changes in angle of attack on both wing and tail.

a. **Aerodynamic pitching moment due to an angle of attack disturbance**

When an aircraft is flying with a steady state angle of attack (e.g., cruise and climb), a disturbance in angle of attack will cause a perturbed-state pitching moment. This reaction is modeled as the dimensional derivative M_α.

If a disturbance causes the aircraft to experience an increase in angle of attack from trim value ($+\Delta\alpha$), a longitudinally dynamically stable aircraft must develop a negative pitching moment ($-\Delta M$) to decrease the angle of attack to return back to the initial trim value (zero or α_o). If the disturbance in angle of attack is positive (increases the angle of attack), the reacting pitching moment is desired to be negative (to reduce the angle of attack). This implies that the derivative M_α should be negative to have a positive contribution in dynamic longitudinal stability.

$$M_\alpha = \frac{\partial M/\partial \alpha}{I_{yy}} < 0 \tag{3.80}$$

The dimensional derivative M_α is obtained [8] by:

$$M_\alpha = \frac{\overline{q}_1 S \overline{c} C_{m_\alpha}}{I_{yy}} \tag{3.81}$$

where non-dimensional derivative C_{m_α} is given in Eq. (3.64).

Two considerable sources for an angle of attack disturbance are wind shear and wind gust. A horizontal wind shear is a variation of the wind vector with changing altitude. The horizontal wind shear is created by the interaction of the horizontal winds in the two parallel air masses. If an aircraft encounters a horizontal wind shear near the ground (e.g., during takeoff and landing), this derivative plays an important role in flight safety.

Example 3.5 The transport aircraft Boeing 747 is flying at 40,000 ft altitude with Mach 0.9 ($U_1 = 871$ fps). The aircraft is in longitudinal trim and has 2.4° of angle of attack.

$$S = 5500\,\text{ft}^2, \text{MAC} = 27.3\,\text{ft}, W_1 = 600,000\,\text{lbf}; C_{m_\alpha} = -1.6\,1/\text{rad}$$

The aircraft experiences a vertical gust, where the angle of attack is increased by 5°. What pitching moment is generated in response to this gust? Is it stabilizing?

Solution
At 40,000 ft altitude, the air density is 0.000587 slug/ft^3.

$$C_m = C_{m_\alpha}\alpha = -1.6 \times \frac{5}{57.3} = -0.14 \tag{3.62}$$

$$\Delta M = \frac{1}{2}\rho V^2 S C_m C = \frac{1}{2} \times 0.000587 \times (871)^2 \times 5500 \times (-0.14) \times 27.3 \tag{1.21}$$

$$\Delta M = -466,800\,\text{lb ft}$$

The pitching moment is negative, so it is stabilizing.

b. Lift due to an angle of attack disturbance

When an aircraft is flying with a steady state angle of attack (e.g., cruise and climb), a disturbance in angle of attack will cause a perturbed-state lift. This reaction is modeled as the dimensional derivative Z_α or the non-dimensional derivative C_{L_α}. The dimensional derivative Z_α is obtained by:

$$Z_\alpha = \frac{-(C_{z_\alpha} + C_{D_1})\bar{q}_1 S}{m} \tag{3.82}$$

where non-dimensional derivative C_{z_α} is obtained by:

$$C_{z_\alpha} = -C_{L_\alpha} - C_{D_1} \tag{3.83}$$

The non-dimensional derivative C_{L_α} is referred to as the aircraft lift curve slope. We may assume that the aircraft lift curve slope (C_{L_α}) is equal to the wing lift curve slope ($C_{L_{\alpha w}}$), so it is obtained [9] by:

$$C_{L_{\alpha w}} = \frac{dC_{L_w}}{d\alpha} = \frac{C_{l_\alpha}}{1 + \frac{C_{l_\alpha}}{\pi AR}} \tag{3.84}$$

where C_{l_α} is the wing airfoil lift curve slope with a theoretical value of 2π 1/rad. The perturbed-state lift due to a disturbance in angle of attack is created due to an increase in angle of attack on both at wing and tail.

c. Drag due to an angle of attack disturbance

When an aircraft is flying with a steady state angle of attack (e.g., cruise), a disturbance in angle of attack will induce a perturbed-state drag. This reaction is modeled as the dimensional derivative X_α, and is obtained [1] by:

$$X_\alpha = -D_\alpha = \frac{-(C_{D_\alpha} - 2C_{L_1})\bar{q}_1 S}{m} \tag{3.85}$$

where non-dimensional derivative C_{D_α} is obtained by:

$$C_{D_\alpha} = \frac{2C_{L_\alpha} C_{L_1}}{\pi ARe} \tag{3.86}$$

where e is the Oswald efficiency factor.

3.5.2 Aircraft Reaction to Disturbance in Pitch Rate

The disturbance in pitch angle and pitch rate will cause three separate and parallel aircraft reactions: (1) Aerodynamic Pitching moment, (2) Lift, (3) Drag. These perturbed-state moment and forces are discussed here. A pitch rate perturbation will induce local changes in angle of attack on both wing and tail.

a. Aerodynamic pitching moment due to a pitch rate disturbance

When an aircraft is flying with a steady state pitch rate (e.g., pull-up) and a steady state pitch angle (e.g., cruise and climb), a disturbance in pitch rate will cause a perturbed-state pitching moment. This reaction is modeled as the dimensional derivative M_q or the non-dimensional derivative C_{m_q}. The most important non-dimensional longitudinal dynamic stability derivative is the rate of change of pitching moment coefficient with respect to disturbance in pitch rate; C_{m_q}.

If a disturbance causes the aircraft to experience an increase in pitch rate from trim value $(+\Delta q)$, a longitudinally dynamically stable aircraft must develop a negative pitching moment $(-\Delta M)$ or a negative pitching moment coefficient $(-\Delta C_m)$ to decrease the pitch rate to return back to the initial trim value (zero or q_o). If the disturbance in pitch rate is positive (increases the pitch angle), the reacting pitching moment is desired to be negative (to reduce the pitch angle). This implies that the derivative C_{m_q} should be negative to have a positive contribution in dynamic longitudinal stability.

$$C_{m_q} = \frac{\partial C_m}{\partial q} \frac{2U_o}{\overline{C}} < 0 \tag{3.87}$$

The dimensional derivative M_q is obtained [8] by:

$$M_q = \frac{\overline{q}_1 S \overline{C}^2 C_{m_q}}{2U_o I_{yy}} \tag{3.88}$$

where non-dimensional derivative C_{m_q} is:

$$C_{m_q} = -2C_{L_{\alpha_h}} \eta_h V_h \frac{l_h}{\overline{C}} \tag{3.89}$$

where l_t is the distance between wing ac to tail ac. The is referred to as the pitch damping derivative.

Two considerable sources for a pitch rate disturbance are wind shear and violent turbulence. If an aircraft encounters a horizontal wind shear near the ground (e.g., during takeoff and landing), this derivative plays an important role in flight safety. A violent turbulence causes a series of sudden changes in pitch rate that, aircraft stability should be able to handle them.

A considerable source for a pitch rate disturbance is horizontal wind shear which is a variation of the wind vector with changing altitude. The horizontal wind shear is created by the interaction of the horizontal winds in the two parallel air masses. If an aircraft encounters a horizontal wind shear near the ground (e.g., during takeoff and landing), this derivative plays an important role in flight safety. If this derivative is not high enough, the aircraft may stall and crash, when encountering a strong wind shear.

b. **Lift due to a pitch rate disturbance**

When an aircraft is flying with a steady state pitch rate (e.g., pull-up) and a steady state pitch angle (e.g., cruise and climb), a disturbance in pitch rate will cause a perturbed-state lift. This reaction is modeled as the dimensional derivative L_q or the non-dimensional derivative C_{L_q}. The dimensional derivative Z_q is obtained by:

$$Z_q = -\frac{\overline{q}_1 S \overline{C} C_{L_q}}{2U_o m} \tag{3.90}$$

where non-dimensional derivative C_{L_q} is the lift coefficient due to pitch rate ($\frac{\partial C_L}{\partial q} \frac{2U_o}{\overline{C}}$):

$$C_{L_q} = 2C_{L_{\alpha_h}} \eta_h V_H \tag{3.91}$$

The perturbed-state lift due to a disturbance in pitch rate is created due to an induced angle of attack (both at wing and tail).

c. **Drag due to a pitch rate disturbance**

When an aircraft is flying with a steady state pitch rate and a steady state pitch angle, a disturbance in pitch rate will induce a perturbed-state drag. This reaction is modeled as the dimensional derivative D_q or the non-dimensional derivative C_{D_q}. The impact of a disturbance in pitch rate on drag is negligible, so:

$$X_q = -D_q = C_{D_q} = 0 \tag{3.92}$$

3.5.3 Aircraft Reaction to a Sudden Change in Altitude

A sudden change in altitude or an altitude perturbation (Δh) is modeled as a plunging motion which represents a rate of change of angle of attack with time; $\dot{\alpha}$. Two considerable sources for a sudden change in altitude are wind shear and violent turbulence. A vertical wind shear is a variation of the wind vector with changing location. The vertical wind shear is created by the interaction of the vertical winds in the two parallel air masses.

Since this disturbance is happening in a unit time, it is determined by Eq. (1.4) and repeated her for convenience we can write:

$$\dot{\alpha} = \frac{\dot{w}}{U_o} \tag{3.93}$$

The disturbance in altitude (angle of attack rate) will cause three separate and parallel aircraft reactions: (1) Aerodynamic Pitching moment, (2) Lift, (3) Drag. These perturbed-state moment and forces are discussed here.

a. Aerodynamic pitching moment due to an altitude disturbance

When an aircraft is cruising with a steady state altitude, a disturbance in altitude will cause a perturbed-state pitching moment. This reaction is modeled as the dimensional derivative $M_{\dot{\alpha}}$ or the non-dimensional derivative $C_{m_{\dot{\alpha}}}$.

If a disturbance causes the aircraft to experience a plunging motion ($+\dot{\alpha}$), a longitudinally dynamically stable aircraft must develop a negative pitching moment ($-\Delta M$) or a negative pitching moment coefficient ($-\Delta C_m$) to nose down to decrease the $\dot{\alpha}$ to return back to the initial trim value (i.e., zero). This implies that the derivative $C_{m_{\dot{\alpha}}}$ should be negative to have a positive contribution in dynamic longitudinal stability.

$$C_{m_{\dot{\alpha}}} = \frac{\partial C_m}{\partial \dot{\alpha}} \frac{2U_o}{\overline{C}} < 0 \tag{3.94}$$

The dimensional derivative M_q is obtained [8] by:

$$M_{\dot{\alpha}} = \frac{\overline{q}_1 S \overline{c}^2 C_{m_{\dot{\alpha}}}}{2 I_{yy} U_o} \tag{3.95}$$

where non-dimensional derivative $C_{m_{\dot{\alpha}}}$ is:

$$C_{m_{\dot{\alpha}}} = -2 C_{L_{\alpha_h}} \eta_h V_H \frac{l_h}{\overline{C}} \frac{d\varepsilon}{d\alpha} \tag{3.96}$$

where l_t is the distance between wing ac to tail ac. If this derivative is not high enough, the aircraft may stall and crash, after a plunging motion.

b. Lift due to an altitude disturbance

When an aircraft is flying with a steady state altitude (e.g., cruise), a disturbance in altitude will cause a perturbed-state lift. This reaction is modeled as the dimensional derivative $Z_{\dot{\alpha}}$ and the non-dimensional derivative $C_{z_{\dot{\alpha}}}$. The dimensional derivative $Z_{\dot{\alpha}}$ is obtained by:

$$Z_{\dot{\alpha}} = \frac{\overline{q}_1 S \overline{C} C_{z_{\dot{\alpha}}}}{2 U_o m} \tag{3.97}$$

where non-dimensional derivative $C_{z_{\dot{\alpha}}}$ is vertical force coefficient due to change in angle of attack $(\frac{\partial C_z}{\partial \dot{\alpha}} \frac{2 U_o}{\overline{C}})$:

$$C_{z_{\dot{\alpha}}} = -C_{L_{\dot{\alpha}}} = -2 C_{L_{\alpha_h}} \eta_h V_h \frac{d\varepsilon}{d\alpha} \tag{3.98}$$

The perturbed-state lift due to a disturbance in altitude is created due to a sudden induced angle of attack (both at wing and tail).

c. **Drag due to an altitude disturbance**

When an aircraft is flying with a steady state altitude, a disturbance in altitude will induce a perturbed-state drag. This reaction is modeled as the dimensional derivative $X_{\dot{\alpha}}$ or the non-dimensional derivative $C_{x_{\dot{\alpha}}}$. The impact of a disturbance in altitude on drag is negligible, so:

$$X_{\dot{\alpha}} = -D_{\dot{\alpha}} = C_{D_{\dot{\alpha}}} = 0 \tag{3.99}$$

3.5.4 Aircraft Reaction to Disturbance in Airspeed

A sudden change in airspeed or an airspeed perturbation (Δu) is experienced, when an aircraft enters a wind shear with velocity profiles. The disturbance in airspeed will cause three separate and parallel aircraft reactions: (1) Aerodynamic Pitching moment, (2) Lift, (3) Drag.

a. **Pitching moment due to an airspeed disturbance**

When an aircraft is cruising with a steady state speed, a disturbance in speed will cause a perturbed-state pitching moment. This reaction is modeled as the dimensional derivative M_u which is obtained by:

$$M_u = \frac{\overline{q}_1 S \overline{C} C_{m_u}}{I_{yy} U_o} \tag{3.100}$$

where non-dimensional derivative C_{m_u} is the pitching moment coefficient due to forward speed derivative $(\frac{\partial C_m}{\partial u} U_o)$:

$$C_{m_u} = -C_{L_1} M_1 \frac{\partial x_{np}}{\partial M} \tag{3.101}$$

The impact of a disturbance in airspeed on pitching moment in subsonic flight regime is negligible. However, for transonic and supersonic speeds, it can be considerable, and may result in unacceptable flying quality. For transonic and supersonic speeds, wind tunnel tests can be employed to estimate this derivative.

b. Lift due to an airspeed disturbance

When an aircraft is flying with a steady state speed (e.g., cruise), a disturbance in airspeed will cause a perturbed-state lift. This reaction is modeled as the dimensional derivative Z_u or the non-dimensional derivative C_{z_u}. The dimensional derivative Z_u is obtained by:

$$Z_u = \frac{-(C_{L_u} + 2C_{L_1})\overline{q}_1 S}{mU_o};$$

(3.102)

where non-dimensional derivative C_{L_u} is the lift coefficient due to change in forward speed ($\frac{\partial C_L}{\partial u} U_o$). For subsonic speeds:

$$C_{L_u} = -C_{z_u} = -C_{L_1} \frac{M_1^2}{1 - M_1^2}$$

(3.103)

For transonic and supersonic speeds, wind tunnel tests can be employed to estimate this derivative.

c. Drag due to an airspeed disturbance

When an aircraft is flying with a steady state airspeed, a disturbance in airspeed will generate a perturbed-state drag. This reaction is modeled as the dimensional derivative X_u or the non-dimensional derivative C_{x_u}. The dimensional derivative X_u is obtained by:

$$X_u = \frac{-(C_{D_u} + 2C_{D_1})\overline{q}_1 S}{mU_o}$$

(3.104)

If a disturbance causes the aircraft to experience an increase in airspeed from trim airspeed ($+\Delta u$), a longitudinally dynamically stable aircraft must develop more drag ($+\Delta D$ or $+\Delta C_D$) to decrease the airspeed to return back to the initial trim value (U_o). Hence, the drag coefficient due to forward speed derivative should be positive for an aircraft to be longitudinally dynamically stable.

$$C_{D_u} = \frac{dC_D}{du} U_o > 0$$

(3.105)

Example 3.6 A business jet aircraft with a maximum take off weight of 12,000 lb, a wing area of 240 ft^2, a wing mean aerodynamic chord of 6 ft is cruising with an airspeed of 200 knot at 20,000 ft altitude. This aircraft has the following longitudinal stability derivatives:

$$C_{m_{\dot{\alpha}}} = -9.2\,\frac{1}{\text{rad}};\; C_{L_{\dot{\alpha}}} = 2.6\,\frac{1}{\text{rad}}$$

Because of a sudden change in the atmospheric pressure, the aircraft is plunging down to a vertical velocity of 12 knot in one second. Calculate the changes in lift and pitching moment of the aircraft right after this incidence. Explain what happens to the aircraft afterward (in terms of upward and pitching motion).

Solution

This plunging motion (Δh) represents a rate of change of angle of attack with time; $\dot{\alpha}$.

$$\alpha \cong \frac{w}{U_o} = \frac{12}{200} = 0.06\,\text{rad} \tag{1.2}$$

Since this motion/disturbance is happening in one second:

$$\dot{\alpha} = \frac{d\alpha}{dt} = \frac{\dot{w}}{U_o} = \frac{0.06}{1} = 0.06\,\frac{\text{rad}}{\text{s}} \tag{1.4}$$

At 20,000 ft altitude, the air density is 0.001267 slug/ft^3. The lift and pitching moment coefficients will be computed by using Eqs. (3.94) and (3.98). However, all perturbations are zero, except $\dot{\alpha}$. thus:

$$C_L = C_{L_{\dot{\alpha}}}\dot{\alpha}\frac{C}{2U_o} = 2.6 \times 0.06 \times \frac{6}{2 \times 200 \times 1.69} = 0.0014 \tag{3.61}$$

$$C_m = C_{m_{\dot{\alpha}}}\dot{\alpha}\frac{C}{2U_o} = -18 \times 0.06 \times \frac{6}{2 \times 200 \times 1.69} = -0.0049 \tag{3.62}$$

The value 1.69 is used to convert knot to ft/s. The lift and pitching moment:

$$\Delta L = \frac{1}{2}\rho V^2 SC_L = \frac{1}{2} \times 0.001267 \times (200 \times 1.69)^2$$
$$\times 240 \times (0.0014) = 24\,\text{lb} \tag{1.19}$$

$$\Delta M = \frac{1}{2}\rho V^2 SC_m C = \frac{1}{2} \times 0.001267$$
$$\times (200 \times 1.69)^2 \times 240 \times (-0.0049) \times 6 \tag{1.21}$$

$$\Delta M = -510\,\text{lb ft}$$

A positive lift (only 24 lb) implies that the aircraft will climb to restore altitude. Moreover, a negative pitching moment (counter-clockwise) indicates that the aircraft will nose down to oppose the disturbance and restore.

3.6 Contributions of Aircraft Components to Longitudinal Stability

Every aircraft component (e.g., wing, fuselage, tail, and propulsion unit) has a unique contribution to the aircraft overall longitudinal stability. An aircraft is longitudinally either stable, or unstable, or neutrally stable; however, a component may have a positive contribution (be stabilizing) or a negative contribution (be destabilizing).

These discussions aim to demonstrate the relationship between the longitudinal stability derivatives and the geometric and configuration characteristics of the aircraft. An important application of this information is during aircraft design process, which helps the designer to select the best configuration to meet longitudinal handling quality requirements.

Main component contribution is explored with respect to the role each play in generating a restoring pitching moment. To analyze aircraft components contributions to longitudinal static stability, we focus on the role of each component (1. sign, and 2. value) in longitudinal static stability derivative, C_{m_α}. In order to analyze aircraft components contributions to longitudinal dynamic stability, we mainly focus on the role of each component (1. positive/negative, and 2. amount) in pitch damping derivative, C_{m_q}.

Equations (3.54) and (3.87) include the primary contributors to longitudinal stability, so for other components, you may refer to Ref. [10]. In this section, contributions of aircraft components to longitudinal stability will be discussed.

3.6.1 Wing Contribution

In Eq. (3.54), the first term represents the main impact of wing-fuselage to the sign and value of static longitudinal stability derivative ($C_{m_{\alpha_{wf}}}$) which is repeated here for convenience.

$$C_{m_{\alpha_{wf}}} = C_{L_{\alpha_{wf}}} \left(\frac{x_{cg}}{\overline{C}} - \frac{x_{ac_{wf}}}{\overline{C}} \right) \tag{3.106}$$

To have a longitudinally statically stable aircraft, the C_{m_α} should be negative. Since $C_{L_{\alpha_{wf}}}$ is always positive (since, the wing is normally constructed of airfoils having a positive camber), the contribution of the wing is a function of the locations of aircraft cg and wing-fuselage ac.

If the aircraft cg is aft of wing-fuselage ac ($x_{cg} > x_{ac_{wf}}$), the wing has a negative contribution which means that it is longitudinally destabilizing. However, If the aircraft cg is ahead of wing-fuselage ac ($x_{cg} < x_{ac_{wf}}$), the wing has a negative contribution which means that it is longitudinally stabilizing. As the distance between aircraft cg and wing ac increases, the impact is more significant, and the contribution is greater.

For a wing-alone configuration aircraft (e.g., hang gliders) to be statically stable, Eq. (3.104) indicates that the wing-fuselage aerodynamic center must lie aft of the aircraft center of gravity. Another wing parameter that impacts its contribution is the downwash slope at the tail which is formulated as $\left(1 - \frac{d\varepsilon}{d\alpha}\right)$ in the second term of Eq. (3.64). The downwash slope is obtained via:

$$\frac{d\varepsilon}{d\alpha} = \frac{2C_{L_{\alpha_w}}}{\pi AR_w} \tag{4.107}$$

The goal is to make this slope as close to zero as possible. From this equation, one can conclude that the higher the wing aspect ratio, the lower is the negative impact of the wing on longitudinal static stability. In another term, to reduce the negative impact of the wing on the positive impact of the tail to static longitudinal stability, increase wing aspect ratio.

The wing aerodynamic center (ac) is moved aft by the sweep angle at about few percent. The aft movement of the ac with increase in sweep angle occurs because the effect of the downwash pattern associated with a swept wing is to raise the lift coefficient on the outer wing panel relative to the inboard lift coefficient.

Since sweep moves the wing outer panel aft relative to the inner portion, the effect on the center of lift is a rearward movement. The effect of wing sweep on ac position is shown in Fig. 3.12 for several combinations of AR and λ.

A fixed-wing aircraft with a conventional wing airfoil (i.e., positive cambered or symmetric) and without a horizontal tail is inherently longitudinally unstable (If cg is aft of ac). If the aircraft is not equipped with an autopilot to provide artificial stability, one solution is to reconfigure the wing to have a reflexed trailing edge. This technique (a wing in

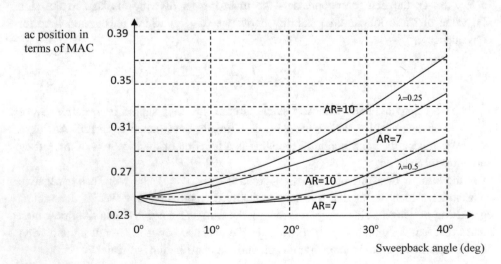

Fig. 3.12 Effect of wing sweep angle on ac position

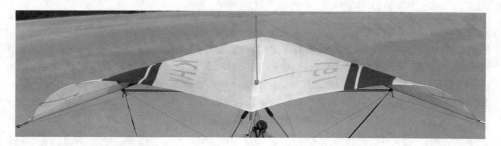

Fig. 3.13 Hang glider Kitty Hawk. By Rob from Cambridge, MA. https://commons.wikimedia.org/w/index.php?curid=2643767

which the camber curves upward toward the trailing edge) is usually employed in flying wing configurations (e.g., Stealth bomber Northrop B-2 Spirit) and hang gliders, such as Kitty Hawk (Fig. 3.13).

3.6.2 Horizontal Tail Contribution

In Eq. (3.64), the second term represents the impact of horizontal tail to the sing and value of static stability derivative ($C_{m_{\alpha_h}}$) which is repeated here for convenience.

$$C_{m_{\alpha_h}} = -C_{L_{\alpha_h}} \eta_h \frac{S_h}{S} \left(\frac{x_{ac_h}}{\overline{C}} - \frac{x_{cg}}{\overline{C}} \right) \left(1 - \frac{d\varepsilon}{d\alpha} \right) \tag{3.108}$$

To have a longitudinally statically stable aircraft, the C_{m_α} should be negative. This term has a negative sign, which implies that an aft horizontal tail is always longitudinally statically stabilizing. There are five terms that represent the amount of tail stabilizing contribution: (1) Tail area (S_h), (2) Tail arm, l_h (i.e., $x_{ac_h} - x_{cg}$), (3) Tail lift curve slope ($C_{L_{\alpha_h}}$), (4) Tail efficiency (η_h), and (5) Downwash slope at the tail ($\frac{d\varepsilon}{d\alpha}$). The first two parameters can be integrated in the tail volume ratio (V_h). The tail lift curve slope can be increased by proper design/selection of its airfoil, plus increasing tail aspect ratio.

For an aircraft with a V-tail or Y-tail (e.g., General Atomics MQ-1 Predator UAV), tail area (S_h) is the horizontal projection of the total planform area of the V-tail (S_{vt}):

$$S_h = S_{vt} \cos(\Gamma_{vt}) \tag{3.109}$$

where Γ_{vt} is the V-tail dihedral angle (viewed from front view). In the Predator UAV, tail dihedral angle of the top portion is estimated to be 32°.

As the tail area, tail arm, tail lift curve slope, and tail efficiency are increased, the tail positive contribution is greater. Moreover, as the downwash slope at the tail is increased, the tail contribution is decreased. To help tail to have a higher positive stabilizing impact, one needs to move the horizontal tail away from the downwash area of the wing (increase its vertical and longitudinal spacing). One example is to change the horizontal tail configuration from conventional to T-tail. The horizontal tail contribution to the static longitudinal stability of the aircraft can be tailored by proper selection of V_h and $C_{L_{\alpha_h}}$.

VTOL and rotary-wing aircraft (e.g., helicopters and quadcopters) are inherently longitudinally unstable, unless they have a fixed horizontal tail. In such case, they require an artificial stability provider (e.g., via an automatic flight control systems) to fly safely.

3.6.3   Vertical Tail Contribution

The vertical tail contribution in a symmetric configuration aircraft has no direct contribution to the static and dynamic longitudinal stability. This is evident by looking at Eqs. (3.54) and (3.87). However, since vertical tail is in the same location of the horizontal tail (in a conventional tail configuration), there is an indirect minor impact.

3.6.4   Fuselage Contribution

The fuselage has an impact of on the wing aerodynamic center location (sometime called Munk effect), which indirectly affects the longitudinal stability. In most cases, the fuselage moves the wing ac forward which means that it makes the aircraft longitudinally more stable. The fuselage is divided into increments (top-views and side-views), the centerlines are connected, and the contribution of all sections are integrated. For the technique on the shift of wing ac due to fuselage, the interested reader may see Part VI of [11].

Another effect of fuselage comes from its cg location which is often about 50% of its length. Due to this feature which moves the aircraft cg aft, the fuselage is longitudinally destabilizing.

3.6.5   Aircraft Weight, Moment of Inertia, and Center of Gravity Contributions

Aircraft weight, aircraft mass moment of inertia about y-axis (i.e., I_{yy}), and aircraft center of gravity have direct contributions to static and dynamic longitudinal stability. Since every component has weight, center of gravity, and geometry; all aircraft components have a direct contribution to static and dynamic longitudinal stability. Any change in weight, center of gravity, and geometry of a component will directly impact longitudinal stability. Hence, one way to improve longitudinal stability is to change the components' locations and the weight distribution.

The location of the aircraft center of gravity has a direct impact to the static and dynamic longitudinal stability. As the cg is moved forward, the aircraft is statically and dynamically longitudinally more stable. The more aft the cg, the component contribution become more destabilizing.

The mass moment of inertia about y-axis (i.e., I_{yy}) is a function of components weight, geometries, and their distance to y-axis. As the aircraft weight and I_{yy} are increased (keeping the same geometry), the aircraft become more longitudinally stable.

3.6.6 Contributions of Other Components

There are more aircraft components which have significant contributions to longitudinal stability such as landing gear and propulsion unit. The actual contribution of the propulsion system and landing gear to the longitudinal stability is much more difficult to estimate. For a propeller driven aircraft, the propeller will develop a normal force in its plane of rotation when the propeller is at an angle of attack. For a jet aircraft, the engine inlet and exhaust nozzle are impacted, when the thrust line is at an angle of attack. In a single engine propeller driven aircraft, the propwash has contribution to the tail efficiency and the downwash field.

If the thrust line is offset from the center of gravity, the propulsive force will create a pitching moment. An engine above the cg will generate a nose down pitching moment, while an engine below the cg will generate a nose up pitching moment. When the aircraft angle of attack is disturbed, these moments will be changed too. The propulsive effects on aircraft stability are commonly estimated from powered wind-tunnel tests. The discussion on these components contributions is beyond the scope of this book.

3.7 Problems

1. Consider a fixed-wing General Aviation aircraft—shown below—with a maximum takeoff weight of 2200 lb, a wing planform area of 162 ft^2, and the following characteristics (Fig. 3.14):

$$\frac{d\varepsilon}{d\alpha} = 0.32; \; C_{L_{\alpha_{wf}}} = 5.7 \, \frac{1}{\text{rad}}; \; C_{L_{\alpha_h}} = 4.3 \, \frac{1}{\text{rad}}; \; b = 31 \, \text{ft}; \; \eta_h = 0.96; \; S_h = 38 \, \text{ft}^2$$

Determine longitudinally static statically derivative; C_{m_α}, neutral point position (np), and static margin (SM). Is this aircraft longitudinally statically stable?

2. Consider a large transport aircraft—shown below—with a maximum takeoff weight of 500,000 lb, a wing planform area of 4000 ft^2, and the following characteristics (Fig. 3.15):

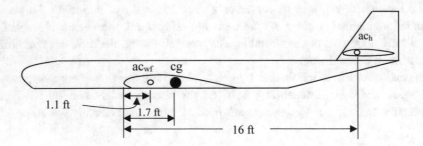

Fig. 3.14 General aviation aircraft for problem 1 (not to scale)

$$\frac{d\varepsilon}{d\alpha} = 0.28; \ C_{L_{\alpha_{wf}}} = 5.5 \ \frac{1}{\text{rad}}; \ C_{L_{\alpha_h}} = 4.5 \ \frac{1}{\text{rad}}; \ b = 200 \, \text{ft}; \ \eta_h = 0.98; \ S_h = 980 \, \text{ft}^2$$

Determine longitudinally static statically derivative; C_{m_α}, neutral point position (np), and static margin (SM). Is this aircraft longitudinally statically stable?

3. A single-engine light aircraft has the following longitudinal characteristic equation for a cruising flight:

$$125s^4 + 308s^3 + 354s^2 + 30s + 29 = 0$$

Is the aircraft dynamically longitudinally stable? Why? If so, determine the features of the longitudinal modes (i.e., damping ratio, natural frequency and period).

4. A fighter aircraft has the following longitudinal characteristic equation for a cruising flight:

$$290s^4 + 267s^3 + 614s^2 + 28s + 11 = 0$$

Is the aircraft dynamically longitudinally stable? Why? If so, determine the features of the longitudinal modes (i.e., damping ratio, natural frequency and period).

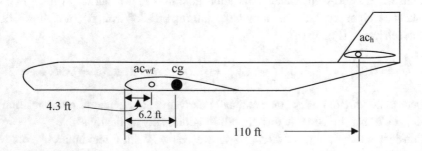

Fig. 3.15 General aviation aircraft for example 2 (not to scale)

5. A GA aircraft has the following angle-of-attack-to-elevator-deflection transfer function for a cruising flight:

$$\frac{\alpha(s)}{\delta_E(s)} = -\frac{28s^3 + 6520s^2 + 3820s + 295}{195s^4 + 2780s^3 + 6200s^2 + 180s + 122}$$

Is the aircraft dynamically longitudinally stable? Why?

6. A GA aircraft has the following pitch-angle-to-elevator-deflection transfer function for a cruising flight:

$$\frac{\theta(s)}{\delta_E(s)} = -\frac{7520s^2 + 13{,}340s + 1180}{243s^4 + 2790s^3 + 5240s^2 + 242s + 178}$$

Is the aircraft dynamically longitudinally stable? Why?

7. A GA aircraft has the following longitudinal state space model for a cruising flight:

$$\begin{bmatrix} \dot{u} \\ \dot{\alpha} \\ \dot{q} \\ \dot{\theta} \end{bmatrix} = \begin{bmatrix} -0.07 & 17 & 0 & -32.2 \\ -0.03 & -1.5 & 0.91 & 0 \\ 0 & -2.5 & -2.7 & 0 \\ 0 & 0 & 1 & 0 \end{bmatrix} \begin{bmatrix} u \\ \alpha \\ q \\ \theta \end{bmatrix} + \begin{bmatrix} 0 & 10 \\ -15 & 0 \\ -5.1 & 0 \\ 0 & 0 \end{bmatrix} \begin{bmatrix} \delta_E \\ \delta_T \end{bmatrix}$$

$$\begin{bmatrix} u \\ \alpha \\ q \\ \gamma \end{bmatrix} = \begin{bmatrix} 1 & 0 & 0 & 0 \\ 0 & 1 & 0 & 0 \\ 0 & 0 & 1 & 0 \\ 0 & -1 & 0 & 1 \end{bmatrix} \begin{bmatrix} u \\ \alpha \\ q \\ \theta \end{bmatrix} + \begin{bmatrix} 0 & 0 \\ 0 & 0 \\ 0 & 0 \\ 0 & 0 \end{bmatrix} \begin{bmatrix} \delta_E \\ \delta_T \end{bmatrix}$$

Is the aircraft dynamically longitudinally stable? Why?

8. A transport aircraft has the following longitudinal state space model for a cruising flight at M = 0.9:

$$\begin{bmatrix} \dot{u} \\ \dot{\alpha} \\ \dot{q} \\ \dot{\theta} \end{bmatrix} = \begin{bmatrix} -0.03 & 22 & 0 & -32.2 \\ -0.07 & -2.3 & 0.85 & 0 \\ 0 & -1.6 & -3.5 & 0 \\ 0 & 0 & 1 & 0 \end{bmatrix} \begin{bmatrix} u \\ \alpha \\ q \\ \theta \end{bmatrix} + \begin{bmatrix} 0 & 14 \\ -25 & 0 \\ -7.2 & 0 \\ 0 & 0 \end{bmatrix} \begin{bmatrix} \delta_E \\ \delta_T \end{bmatrix}$$

$$\begin{bmatrix} u \\ \alpha \\ q \\ \gamma \end{bmatrix} = \begin{bmatrix} 1 & 0 & 0 & 0 \\ 0 & 1 & 0 & 0 \\ 0 & 0 & 1 & 0 \\ 0 & -1 & 0 & 1 \end{bmatrix} \begin{bmatrix} u \\ \alpha \\ q \\ \theta \end{bmatrix} + \begin{bmatrix} 0 & 0 \\ 0 & 0 \\ 0 & 0 \\ 0 & 0 \end{bmatrix} \begin{bmatrix} \delta_E \\ \delta_T \end{bmatrix}$$

Is the aircraft dynamically longitudinally stable? Why?

9. A large transport aircraft has the following speed-to-elevator-deflection transfer function for a cruising flight:

$$\frac{u(s)}{\delta_E(s)} = \frac{-18s^2 + 3780s + 15{,}600}{794s^4 + 967s^3 + 1320s^2 + 184s + 172}$$

If the aircraft is dynamically longitudinally stable, determine features of phugoid and short period modes.

10. A GA aircraft has the following pitch-angle-to-elevator-deflection transfer function for a cruising flight:

$$\frac{\theta(s)}{\delta_E(s)} = \frac{N(s)}{118s^4 + 333s^3 + 2660s^2 + 110s + 35}$$

If the aircraft is dynamically longitudinally stable, determine features of phugoid and short period modes.

11. A twin-engine business jet aircraft has the following angle-of-attack-to-elevator-deflection transfer function for a cruising flight:

$$\frac{\alpha(s)}{\delta_E(s)} = -\frac{8s^3 + 520s^2 + 32s + 41}{183s^4 + 290s^3 + 440s^2 + 24s + 18}$$

If the aircraft is dynamically longitudinally stable, determine features of phugoid and short period modes.

12. A business jet aircraft with a maximum take off weight of 15,000 lb, a wing area of 290 ft^2, a wing mean aerodynamic chord of 7.3 ft is cruising with an airspeed of 240 knot at 20,000 ft altitude. This aircraft has the following longitudinal stability derivatives:

$$C_{m_{\dot{\alpha}}} = -8.4 \frac{1}{\text{rad}}; \, C_{L_{\dot{\alpha}}} = 2.3 \frac{1}{\text{rad}}$$

Because of a sudden change in the atmospheric pressure, the aircraft is plunging down to a vertical velocity of 15 knot in one second. Calculate the changes in lift and pitching moment of the aircraft right after this incidence. Explain what happens to the aircraft afterward (in terms of upward and pitching motion).

13. A large transport aircraft with a maximum take off weight of 700,000 lb, a wing area of 4700 ft^2, a wing mean aerodynamic chord of 22 ft is cruising with an airspeed of 450 knot at 35,000 ft altitude. This aircraft has the following longitudinal stability derivatives:

$$C_{m_{\dot{\alpha}}} = -4.2 \frac{1}{\text{rad}}; \, C_{L_{\dot{\alpha}}} = 7 \frac{1}{\text{rad}}$$

Because of a sudden change in the atmospheric pressure, the aircraft is plunging down to a vertical velocity of 25 knot in one second. Calculate the changes in lift and pitching moment of the aircraft right after this incidence. Explain what happens to the aircraft afterward (in terms of upward and pitching motion).

14. A large transport aircraft is flying at 40,000 ft altitude with an airspeed of 840 fps. The aircraft is in longitudinal trim and has 3° of angle of attack.

$$S = 5000\,\text{ft}^2,\ \text{MAC} = 26\,\text{ft},\ C_{m_\alpha} = -1.8\ 1/\text{rad}$$

The aircraft experiences a vertical gust, where the angle of attack is increased by 3°. What pitching moment is generated in response to this gust? Is it stabilizing?

References

1. Roskam J., Airplane flight dynamics and automatic flight controls, 2007, DARCO
2. Tischler M. B., Remple R. K., *Aircraft and Rotorcraft System Identification*, AIAA education series, Reston, AIAA 2006
3. Jategaonkar, R., D. Fischenberg and W. Von Gruenhagen, Aerodynamic modeling and system identification from flight data - Recent applications at DLR. *Journal of Aircraft*, 2004, 41(4): 681–691
4. De Jesus-Mota, S., M. N. Beaulieu and R. M. Botez, Identification of a MIMO state space model of an F/A-18 aircraft using a subspace method. *Aeronautical Journal*, 2009, 113(1141): 183–190
5. Burri M, Nikolic J, Oleynikova H, Achtelik MW, Siegwart R., Maximum likelihood parameter identification for MAVs, In *2016 IEEE International, Conference on Robotics and Automation*, Piscataway, NJ, IEEE; 4297–4303, 2016
6. Kimberlin R. D., Flight Testing of Fixed Wing Aircraft, AIAA, 2003
7. MIL-STD-1797, Flying Qualities of Piloted Aircraft, Department of Defense, Washington DC, 1997
8. Stevens B. L., Lewis F. L., E. N. Johnson, Aircraft control and simulation, Third edition, John Wiley, 2016
9. Anderson J. D., Fundamentals of Aerodynamics, McGraw-Hill, Sixth edition, 2016
10. Hoak D. E., Ellison D. E., et al, USAF Stability and Control DATCOM, Flight Control Division, Air Force Flight Dynamics Laboratory, Wright-Patterson AFB, Ohio, 1978
11. Roskam J., Airplane Design, 2015, DAR Corp

Lateral-Directional Stability

4

4.1 Fundamentals

Definition, fundamentals, and categories of stability were presented in Chap. 3. Three basic axes are: (1) Lateral stability (about x axis), (2) Longitudinal stability (about y axis), and (3) Directional stability (about z axis). As discussed in Chap. 1, we treated longitudinal stability independent of lateral-directional stability, and longitudinal motions are decoupled from lateral-directional motions. Then, the independent longitudinal state equations and lateral-directional state equations are derived. Moreover, there is a strong coupling between lateral and directional motions and state variables. Any lateral moment—which is to generate a lateral motion—often produces a directional motion. On the other hand, any directional moment—which is to generate a directional motion—often produces a lateral motion.

Lateral-directional stability refers to the tendency of an aircraft to oppose any lateral-directional disturbance, and to return to its lateral-directional equilibrium state, if disturbed. The lateral-directional stability is inherent and, primarily provided by the air vehicle configuration. When the air vehicle configuration is changed, its lateral-directional stability characteristics will vary.

We categorize lateral-directional stability in two modes: (1) Lateral-directional static stability, and (2) Lateral-directional dynamic stability. Lateral-directional static stability refers to the tendency of an aircraft to develop forces and moments to oppose any lateral-directional disturbance/perturbations. Lateral-directional dynamic stability dynamic stability is to return to the present lateral-directional trim condition, if disturbed.

Multiple atmospheric disturbances (see Fig. 4.1) may apply perturbations to the lateral-directional flying motion. For instance: (1) Horizontal gust when hits on one side of the fuselage nose, or rear fuselage, or the vertical tail, (2) Vertical gust when hits under one side of wing or the horizontal tail, and (3) Turbulence. A flying aircraft must be

© The Author(s), under exclusive license to Springer Nature Switzerland AG 2022 109
M. H. Sadraey, *Flight Stability and Control*, Synthesis Lectures on Mechanical
Engineering, https://doi.org/10.1007/978-3-031-18765-0_4

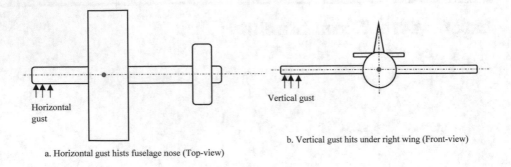

a. Horizontal gust hists fuselage nose (Top-view)

b. Vertical gust hits under right wing (Front-view)

Fig. 4.1 Two sample atmospheric lateral-directional perturbations

stable, and maintain trim conditions, when facing these disturbing/hazardous atmospheric phenomena.

In general, most fixed-wing air vehicles are often laterally-directionally stable (e.g., Cessna 172), while VTOL aircraft including helicopters (e.g., DJI Phantom) are inherently laterally-directionally unstable, unless they have a fixed vertical tail (e.g., McDonnell Douglas Helicopter 520 N NOTAR). In VTOL aircraft, the stability is often artificially provided by the automatic flight control systems. Some large transport aircraft are either laterally lightly stable, or even laterally unstable. In large transport aircraft (e.g., Boeing 767 and Airbus A380), lateral-directional stability is often artificially augmented by autopilot devices such as yaw damper and roll damper.

An aircraft designer should consider the lateral-directional stability requirements, and shall make sure that, the aircraft is laterally-directionally stable within the design flight envelope. In this chapter, static and dynamic lateral-directional stability of an aircraft are examined, and their requirements and analysis techniques are presented. Moreover, the aircraft reactions to lateral-directional perturbations and aircraft components contributions to lateral-directional stability are discussed. Before analysis techniques are discussed, we need to derive lateral-directional motions governing equations.

4.2   Lateral-Directional Governing Equations of Motion

4.2.1   Governing Differential Equations of Motion

The nonlinear six-degrees of freedom aircraft governing equations of motion are derived in Chap. 1 (Eqs. 1.28–1.33). In Chap. 2, we have decoupled longitudinal motions from lateral-directional equations. Now, we collect the lateral-directional equations of motion as a group. Three degrees of freedom features a lateral-directional motion: (1) Linear motion along y-axis, (2) Rotational motion about x-axis (i.e., roll), (3) Rotational motion about z-axis (i.e., yaw). To analyze lateral-directional stability, we need to derive lateral-directional transfer functions and/or lateral-directional state space model.

For linear motion along y-axis, variables are linear displacement (to the left or right, y), linear speed (v), and linear acceleration ($\dot{v}$). For angular motion about x-axis, variables are angular displacement (bank angle, ϕ), angular speed or roll rate (p), and angular acceleration ($\dot{p}$). For angular motion about z-axis, variables are angular displacement (yaw and heading angle, ψ), angular speed or yaw rate (r), and angular acceleration ($\dot{r}$).

One force Eq. (1.44), and two moment Eqs. (1.46) and (1.48) govern the lateral-directional motion (in yz and xy planes). The equations are repeated here for convenience:

$$m(\dot{v} + U_1 r + R_1 u - W_1 p - P_1 w) = f_{A_Y} + f_{T_Y}$$
$$+ mg(\cos\Phi_1\cos\Theta_1\phi - \sin\Phi_1\sin\Theta_1\theta) \tag{4.1}$$

$$I_{xx}\dot{p} - I_{xz}\dot{r} - I_{xz}(P_1 q + Q_1 p) + (I_{zz} - I_{yy})(R_1 q - Q_1 r) = l_A - l_T \tag{4.2}$$

$$I_{zz}\dot{r} - I_{xz}\dot{p} + (I_{yy} - I_{xx})(P_1 q + Q_1 p) + I_{xz}(Q_1 r + R_1 q) = n_A + n_T \tag{4.3}$$

To analyze lateral-directional stability, we need to isolate lateral-directional motions from longitudinal motions. Hence, we drop longitudinal flight parameters (i.e., $q = \theta = w = 0$) from lateral-directional governing equations of motion. Moreover, a level cruising (steady state symmetric straight line) flight is considered, so $W_1 = P_1 = R_1 = \Phi_1 = 0$. Furthermore, it is assumed that engine thrust is symmetric along x-axis and will not be impacted by any lateral-directional disturbance (i.e., $n_T = l_T = f_{T_y} = 0$). For aerodynamic forces along y, and rolling and yawing moments, the new symbols Y, Z and N are utilized respectively. Thus, the equations are simplified to the following form:

$$m(\dot{v} + U_1 r) = Y + mg(\cos\Theta_1\phi) \tag{4.4}$$

$$I_{xx}\dot{p} - I_{xz}\dot{r} = L_A \tag{4.5}$$

$$I_{zz}\dot{r} - I_{xz}\dot{p} = N \tag{4.6}$$

Next step is to employ linearized side force, or crosswind force (Y) and rolling and yawing moments (L_A and N), as derived in Chap. 1. One directional force and two lateral-directional moments equations (Eqs. 1.54, 1.55, and 1.57) are repeated here for convenience.

$$Y = \frac{\partial Y}{\partial\beta}\beta + \frac{\partial F_Y}{\partial\dot{\beta}}\dot{\beta} + \frac{\partial F_Y}{\partial p}p + \frac{\partial F_Y}{\partial r}r + \frac{\partial F_Y}{\partial\delta_A}\delta_A + \frac{\partial F_Y}{\partial\delta_R}\delta_R \tag{4.7}$$

$$L = \frac{\partial L}{\partial\beta}\beta + \frac{\partial L}{\partial p}p + \frac{\partial L}{\partial r}r + \frac{\partial L}{\partial\delta_A}\delta_A + \frac{\partial L}{\partial\delta_R}\delta_R \tag{4.8}$$

$$N = \frac{\partial N}{\partial \beta}\beta + \frac{\partial N}{\partial p}p + \frac{\partial N}{\partial r}r + \frac{\partial N}{\partial \delta_A}\delta_A + \frac{\partial N}{\partial \delta_R}\delta_R \tag{4.9}$$

The aerodynamic side force and rolling and yawing moment can be expressed as a function of all the motion variables. However, in these linear equations, only the terms that are significant have been retained. To keep the format of Eqs. (4.4)–(4.6), we use new parameters for dimensional derivatives:

$$Y_\beta = \frac{\partial Y/\partial \beta}{m}; \; Y_p = \frac{\partial Y/\partial p}{m}; \; Y_r = \frac{\partial Y/\partial r}{m}; \; Y_{\delta_A} = \frac{\partial Y/\partial \delta_A}{m}; \; Y_{\delta_R} = \frac{\partial Y/\partial \delta_R}{m} \tag{4.10}$$

$$L_\beta = \frac{\partial L/\partial \beta}{I_{xx}}; \; L_p = \frac{\partial L/\partial p}{I_{xx}}; \; L_r = \frac{\partial L/\partial r}{I_{xx}}; \; L_{\delta_A} = \frac{\partial L/\partial \delta_A}{I_{xx}}; \; L_{\delta_R} = \frac{\partial L/\partial \delta_R}{I_{xx}} \tag{4.11}$$

$$N_\beta = \frac{\partial N/\partial \beta}{I_{zz}}; \; N_p = \frac{\partial N/\partial p}{I_{zz}}; \; N_r = \frac{\partial N/\partial r}{I_{zz}}; \; N_{\delta_A} = \frac{\partial N/\partial \delta_A}{I_{zz}}; \; N_{\delta_R} = \frac{\partial N/\partial \delta_R}{I_{zz}} \tag{4.12}$$

Hence, the linear governing equations for lateral-directional motions will be reformatted as:

$$\dot{v} + U_1 r = g\phi(\cos \Theta_1) + Y_\beta \beta + Y_p p + Y_r r + Y_{\delta_A}\delta_A + Y_{\delta_R}\delta_R \tag{4.13}$$

$$\dot{p} - \frac{I_{xz}}{I_{xx}}\dot{r} = L_\beta \beta + L_p p + L_r r + L_{\delta_A}\delta_A + L_{\delta_R}\delta_R \tag{4.14}$$

$$\dot{r} - \frac{I_{xz}}{I_{zz}}\dot{p} = N_\beta \beta + N_p p + N_r r + N_{\delta_A}\delta_A + N_{\delta_R}\delta_R \tag{4.15}$$

The dimensional derivatives are determined from the following equations:
Aerodynamic side force derivatives along y-axis:

$$Y_\beta = \frac{-\overline{q}_1 S C_{y_\beta}}{m}; \; Y_p = \frac{\overline{q}_1 S b C_{y_p}}{2mU_1}; \; Y_r = \frac{\overline{q}_1 S b C_{y_r}}{2mU_1};$$

$$Y_{\delta_A} = \frac{\overline{q}_1 S C_{y_{\delta_A}}}{m}; \; Y_{\delta_R} = \frac{\overline{q}_1 S C_{y_{\delta_R}}}{m} \tag{4.16}$$

Aerodynamic rolling moment derivatives (about x-axis):

$$L_\beta = \frac{\overline{q}_1 S b C_{l_\beta}}{I_{xx}}; \; L_p = \frac{\overline{q}_1 S b^2 C_{l_p}}{2I_{xx}U_1}; \; L_r = \frac{\overline{q}_1 S b^2 C_{l_r}}{2I_{xx}U_1};$$

$$L_{\delta_A} = \frac{\overline{q}_1 S b C_{l_{\delta_A}}}{I_{xx}}; \; L_{\delta_R} = \frac{\overline{q}_1 S b C_{l_{\delta_R}}}{I_{xx}} \tag{4.17}$$

Aerodynamic yawing moment derivatives (about z-axis):

$$N_\beta = \frac{\overline{q}_1 Sb C_{n_\beta}}{I_{zz}}; \ N_p = \frac{\overline{q}_1 Sb^2 C_{n_p}}{2 I_{zz} U_1}; \ N_r = \frac{\overline{q}_1 Sb^2 C_{n_r}}{2 I_{zz} U_1};$$

$$N_{\delta_A} = \frac{\overline{q}_1 Sb C_{n_{\delta_A}}}{I_{zz}}; \ N_{\delta_R} = \frac{\overline{q}_1 Sb C_{n_{\delta_R}}}{I_{zz}} \tag{4.18}$$

We are interested in observing variations of the following motion variables: (1) Sideslip angle (β), (2) Bank angle (ϕ), (3) Yaw angle (ψ). Two primary control inputs to lateral-directional motions are: (1) Aileron deflection (δ_A), and (2) Rudder deflection (δ_R). In a pure rolling motion, the roll rate is

$$p = \dot{\phi} => \dot{p} = \ddot{\phi} \tag{4.19}$$

In a pure yawing motion, the yaw rate is:

$$r = \dot{\psi} => \dot{r} = \ddot{\psi} \tag{4.20}$$

Furthermore, as sideslip angle (β) is defined in Chap. 1:

$$v = U_o \beta => \dot{v} = U_o \dot{\beta} \tag{4.21}$$

Hence, Eqs. (4.13)–(4.15) will be rewritten as:

$$U_o \dot{\beta} + U_o \dot{\psi} = g\phi(\cos \Theta_o) + Y_\beta \beta + Y_p \dot{\phi} + Y_r \dot{\psi} + Y_{\delta_A} \delta_A + Y_{\delta_R} \delta_R \tag{4.22}$$

$$\ddot{\phi} - \frac{I_{xz}}{I_{xx}} \ddot{\psi} = L_\beta \beta + L_p \dot{\phi} + L_r \dot{\psi} + L_{\delta_A} \delta_A + L_{\delta_R} \delta_R \tag{4.23}$$

$$\ddot{\psi} - \frac{I_{xz}}{I_{zz}} \ddot{\phi} = N_\beta \beta + N_p \dot{\phi} + N_r \dot{\psi} + N_{\delta_A} \delta_A + N_{\delta_R} \delta_R \tag{4.24}$$

These linear differential equations govern the lateral-directional motions and can be readily presented into two forms: (1) Transfer function, (2) State space model. In general, classical control simply employs transfer functions (s-domain), while modern control is often based on state-space (time-domain) models. These two lateral-directional dynamic models are derived in the following two sections.

4.2.2 Lateral-Directional Transfer Functions

To derive lateral-directional transfer functions, we first need to apply Laplace transform (L) to all three differential Eqs. (4.22)–(4.24), assuming zero initial conditions. The Laplace transform of first derivative and second derivative are:

$$L(\dot{x}(t)) = sX(s) \tag{4.25}$$

and

$$L(\ddot{x}(t)) = s^2 X(s) \tag{4.26}$$

Applying these two transforms to Eqs. (4.22)–(4.24) yields:

$$U_o s\beta(s) + U_o s\psi(s) = g\cos\Theta_o\phi(s) + Y_\beta\beta(s) + Y_p s\phi(s)$$
$$+ Y_r s\psi(s) + Y_{\delta_A}\delta_A(s) + Y_{\delta_R}\delta_R(s) \tag{4.27}$$

$$s^2\phi(s) - \frac{I_{xz}}{I_{xx}}s^2\psi(s) = L_\beta\beta(s) + L_p s\phi(s) + L_r s\psi(s)$$
$$+ L_{\delta_A}\delta_A(s) + L_{\delta_R}\delta_R(s) \tag{4.28}$$

$$s^2\psi(s) - \frac{I_{xz}}{I_{zz}}s^2\phi(s) = N_\beta\beta(s) + N_p s\phi(s) + N_r s\psi(s) + N_{\delta_A}\delta_A(s) + N_{\delta_R}\delta_R(s) \tag{4.29}$$

Then, these equations are reformatted as factors of two inputs $(\delta_A(s), \delta_R(s))$ and three outputs $(\beta(s), \phi(s), \psi(s))$:

$$\left[sU_o - Y_\beta\right]\beta(s) - \left[sY_p + g\cos\Theta_o\right]\phi(s) + [sU_o - sY_r]\psi(s) = Y_{\delta_A}\delta_A(s) + Y_{\delta_R}\delta_R(s) \tag{4.30}$$

$$\left[-L_\beta\right]\beta(s) + \left[s^2 - sL_p\right]\phi(s) - \left[\frac{I_{xz}}{I_{xx}}s^2 + L_r s\right]\psi(s) = L_{\delta_A}\delta_A(s) + L_{\delta_R}\delta_R(s) \tag{4.31}$$

$$\left[-N_\beta\right]\beta(s) - \left[\frac{I_{xz}}{I_{zz}}s^2 + sN_p\right]\phi(s) + \left[s^2 - sN_r\right]\psi(s) = N_{\delta_A}\delta_A(s) + N_{\delta_R}\delta_R(s) \tag{4.32}$$

Writing these three equations in matrix format will allows us to use Cramer's rule to derive lateral-directional transfer functions. For now, we are to derive six transfer functions: $\frac{\beta(s)}{\delta_A(s)}$, $\frac{\phi(s)}{\delta_A(s)}$, $\frac{\psi(s)}{\delta_A(s)}$, $\frac{\beta(s)}{\delta_R(s)}$, $\frac{\phi(s)}{\delta_R(s)}$, and $\frac{\psi(s)}{\delta_R(s)}$. For aileron deflection (δ_A) input, we obtain:

$$\begin{bmatrix} sU_o - Y_\beta & -\left[sY_p + g\cos\Theta_o\right] & sU_o - sY_r \\ -L_\beta & s^2 - sL_p & -\left[\frac{I_{xz}}{I_{xx}}s^2 + L_r s\right] \\ -N_\beta & -\left[\frac{I_{xz}}{I_{zz}}s^2 + sN_p\right] & s^2 - sN_r \end{bmatrix}\begin{bmatrix} \frac{\beta(s)}{\delta_A(s)} \\ \frac{\phi(s)}{\delta_A(s)} \\ \frac{\psi(s)}{\delta_A(s)} \end{bmatrix} = \begin{bmatrix} Y_{\delta_A} \\ L_{\delta_A} \\ N_{\delta_A} \end{bmatrix} \tag{4.33}$$

For rudder deflection (δ_R) input, we obtain:

$$\begin{bmatrix} sU_o - Y_\beta & -\left[sY_p + g\cos\Theta_o\right] & sU_o - sY_r \\ -L_\beta & s^2 - sL_p & -\left[\frac{I_{xz}}{I_{xx}}s^2 + L_r s\right] \\ -N_\beta & -\left[\frac{I_{xz}}{I_{zz}}s^2 + sN_p\right] & s^2 - sN_r \end{bmatrix}\begin{bmatrix} \frac{\beta(s)}{\delta_R(s)} \\ \frac{\phi(s)}{\delta_R(s)} \\ \frac{\psi(s)}{\delta_R(s)} \end{bmatrix} = \begin{bmatrix} Y_{\delta_R} \\ L_{\delta_R} \\ N_{\delta_R} \end{bmatrix} \tag{4.34}$$

For the first unknown (i.e., the sideslip-angle-to-aileron-deflection transfer function; $\frac{\beta(s)}{\delta_A(s)}$); in the numerator, the first column of the coefficient matrix is replaced by the by the column vector of right-hand-side of Eq. (4.33).

$$\frac{\beta(s)}{\delta_A(s)} = \frac{\left\| \begin{bmatrix} Y_{\delta_A} & -\left[sY_p + g\cos\Theta_o\right] & sU_o - sY_r \\ L_{\delta_A} & s^2 - sL_p & -\left[\frac{I_{xz}}{I_{xx}}s^2 + L_r s\right] \\ N_{\delta_A} & -\left[\frac{I_{xz}}{I_{zz}}s^2 + sN_p\right] & s^2 - sN_r \end{bmatrix} \right\|}{\left\| \begin{bmatrix} sU_o - Y_\beta & -\left[sY_p + g\cos\Theta_o\right] & sU_o - sY_r \\ -L_\beta & s^2 - sL_p & -\left[\frac{I_{xz}}{I_{xx}}s^2 + L_r s\right] \\ -N_\beta & -\left[\frac{I_{xz}}{I_{zz}}s^2 + sN_p\right] & s^2 - sN_r \end{bmatrix} \right\|} \tag{4.35}$$

Similarly, for the second unknown (i.e., the bank-angle-to-aileron-deflection transfer function); in the numerator, the second column of the coefficient matrix is replaced by the by the column vector of right-hand-side of Eq. (4.33).

$$\frac{\phi(s)}{\delta_A(s)} = \frac{\left\| \begin{bmatrix} sU_o - Y_\beta & Y_{\delta_A} & sU_o - sY_r \\ -L_\beta & L_{\delta_A} & -\left[\frac{I_{xz}}{I_{xx}}s^2 + L_r s\right] \\ -N_\beta & N_{\delta_A} & s^2 - sN_r \end{bmatrix} \right\|}{\left\| \begin{bmatrix} sU_o - Y_\beta & -\left[sY_p + g\cos\Theta_o\right] & sU_o - sY_r \\ -L_\beta & s^2 - sL_p & -\left[\frac{I_{xz}}{I_{xx}}s^2 + L_r s\right] \\ -N_\beta & -\left[\frac{I_{xz}}{I_{zz}}s^2 + sN_p\right] & s^2 - sN_r \end{bmatrix} \right\|} \tag{4.36}$$

For the third unknown (i.e., the yaw-angle-to-aileron-deflection transfer function); in the numerator, the third column of the coefficient matrix is replaced by the by the column vector of right-hand-side of Eq. (4.33).

$$\frac{\psi(s)}{\delta_A(s)} = \frac{\left\| \begin{bmatrix} sU_o - Y_\beta & -\left[sY_p + g\cos\theta_o\right] & Y_{\delta_A} \\ -L_\beta & s^2 - sL_p & L_{\delta_A} \\ -N_\beta & -\left[\frac{I_{xz}}{I_{zz}}s^2 + sN_p\right] & N_{\delta_A} \end{bmatrix} \right\|}{\left\| \begin{bmatrix} sU_o - Y_\beta & -\left[sY_p + g\cos\theta_o\right] & sU_o - sY_r \\ -L_\beta & s^2 - sL_p & -\left[\frac{I_{xz}}{I_{xx}}s^2 + L_r s\right] \\ -N_\beta & -\left[\frac{I_{xz}}{I_{zz}}s^2 + sN_p\right] & s^2 - sN_r \end{bmatrix} \right\|} \tag{4.37}$$

A similar technique is employed for deriving transfer functions due to rudder deflection (δ_R) input. The only difference is that the base matrix equation is (4.34).

$$\frac{\beta(s)}{\delta_R(s)} = \frac{\left|\left| \begin{array}{ccc} Y_{\delta_R} & -[sY_p + g\cos\theta_o] & sU_o - sY_r \\ L_{\delta_R} & s^2 - sL_p & -\left[\frac{I_{xz}}{I_{xx}}s^2 + L_rs\right] \\ N_{\delta_R} & -\left[\frac{I_{xz}}{I_{zz}}s^2 + sN_p\right] & s^2 - sN_r \end{array} \right|\right|}{\left|\left| \begin{array}{ccc} sU_o - Y_\beta & -[sY_p + g\cos\theta_o] & sU_o - sY_r \\ -L_\beta & s^2 - sL_p & -\left[\frac{I_{xz}}{I_{xx}}s^2 + L_rs\right] \\ -N_\beta & -\left[\frac{I_{xz}}{I_{zz}}s^2 + sN_p\right] & s^2 - sN_r \end{array} \right|\right|} \tag{4.38}$$

$$\frac{\phi(s)}{\delta_R(s)} = \frac{\left|\left| \begin{array}{ccc} sU_o - Y_\beta & Y_{\delta_R} & sU_o - sY_r \\ -L_\beta & L_{\delta_R} & -\left[\frac{I_{xz}}{I_{xx}}s^2 + L_rs\right] \\ -N_\beta & N_{\delta_R} & s^2 - sN_r \end{array} \right|\right|}{\left|\left| \begin{array}{ccc} sU_o - Y_\beta & -[sY_p + g\cos\theta_o] & sU_o - sY_r \\ -L_\beta & s^2 - sL_p & -\left[\frac{I_{xz}}{I_{xx}}s^2 + L_rs\right] \\ -N_\beta & -\left[\frac{I_{xz}}{I_{ZZ}}s^2 + sN_p\right] & s^2 - sN_r \end{array} \right|\right|} \tag{4.39}$$

$$\frac{\psi(s)}{\delta_R(s)} = \frac{\left|\left| \begin{array}{ccc} sU_o - Y_\beta & -[sY_p + g\cos\theta_o] & Y_{\delta_R} \\ -L_\beta & s^2 - sL_p & L_{\delta_R} \\ -N_\beta & -\left[\frac{I_{XZ}}{I_{zz}}s^2 + sN_p\right] & N_{\delta_R} \end{array} \right|\right|}{\left|\left| \begin{array}{ccc} sU_o - Y_\beta & -[sY_p + g\cos\theta_o] & sU_o - sY_r \\ -L_\beta & s^2 - sL_p & -\left[\frac{I_{xZ}}{I_{xx}}s^2 + L_rs\right] \\ -N_\beta & -\left[\frac{I_{xz}}{I_{ZZ}}s^2 + sN_p\right] & s^2 - sN_r \end{array} \right|\right|} \tag{4.40}$$

Now, we need to derive 12 determinants (six for numerators, and six for denominators). By employing matrix algebra and expanding the coefficients, six transfer functions will be obtained.

For aileron deflection input (δ_A):

$$\frac{\beta(s)}{\delta_A(s)} = \frac{s\left[A_{\beta A}s^3 + B_{\beta A}s^2 + C_{\beta A}s + D_{\beta A}\right]}{s\left[A_2s^4 + B_2s^3 + C_2s^2 + D_2s + E_2\right]} \tag{4.41}$$

$$\frac{\phi(s)}{\delta_A(s)} = \frac{s\left[A_{\phi A}s^2 + B_{\phi A}s + C_{\phi A}\right]}{s\left[A_2s^4 + B_2s^3 + C_2s^2 + D_2s + E_2\right]} \tag{4.42}$$

$$\frac{\psi(s)}{\delta_A(s)} = \frac{A_{\psi A}s^3 + B_{\psi A}s^2 + C_{\psi A}s + D_{\psi A}}{s\left[A_2s^4 + B_2s^3 + C_2s^2 + D_2s + E_2\right]} \tag{4.43}$$

For rudder deflection input (δ_R):

$$\frac{\beta(s)}{\delta_R(s)} = \frac{s\left[A_{\beta R}s^3 + B_{\beta R}s^2 + C_{\beta R}s + D_{\beta R}\right]}{s\left[A_2s^4 + B_2s^3 + C_2s^2 + D_2s + E_2\right]} \tag{4.44}$$

$$\frac{\phi(s)}{\delta_R(s)} = \frac{s\left[A_{\phi R}s^2 + B_{\phi R}s + C_{\phi R}\right]}{s\left[A_2 s^4 + B_2 s^3 + C_2 s^2 + D_2 s + E_2\right]} \tag{4.45}$$

$$\frac{\psi(s)}{\delta_R(s)} = \frac{A_{\psi R}s^3 + B_{\psi R}s^2 + C_{\psi R}s + D_{\psi R}}{s\left[A_2 s^4 + B_2 s^3 + C_2 s^2 + D_2 s + E_2\right]} \tag{4.46}$$

Since, in this chapter, we are only interested in analyzing stability, we only concentrate on the denominator. All coefficients of numerators (e.g., $A_{\beta R}$ and $B_{\psi A}$) of lateral-directional transfer functions (i.e., Eqs. 4.41–4.46) will be derived in Chap. 6.

Six lateral-directional transfer functions have distinct numerators, but the same denominator (i.e., determinant of Eqs. (4.41–4.46):

$$TF_{lat-dir} = \frac{Numerators}{A_2 s^5 + B_2 s^4 + C_2 s^3 + D_2 s^2 + E_2 s} \tag{4.47}$$

The lateral-directional characteristic equation is:

$$A_2 s^5 + B_2 s^4 + C_2 s^3 + D_2 s^2 + E_2 s = s\left[A_2 s^4 + B_2 s^3 + C_2 s^2 + D_2 s + E_2\right] = 0 \tag{4.48}$$

This is a fifth-order polynomial in s-domain, with a free s (i.e., s = 0). The coefficients are functions of the airspeed, stability derivatives, and inertia characteristics of the aircraft:

$$A_2 = U_o\left(1 - \frac{I_{xz}}{I_{xx}}\frac{I_{xz}}{I_{zz}}\right) \tag{4.49}$$

$$B_2 = -Y_\beta\left(1 - \frac{I_{xz}}{I_{xx}}\frac{I_{xz}}{I_{zz}}\right) - U_o\left(L_p + N_r + N_p\frac{I_{xz}}{I_{xx}} + L_r\frac{I_{xz}}{I_{zz}}\right) \tag{4.50}$$

$$C_2 = U_o\left(L_p N_r - L_r N_p\right) + Y_\beta\left(L_p + N_r + \frac{I_{xz}}{I_{xx}}N_p + \frac{I_{xz}}{I_{zz}}L_r\right)$$
$$- Y_p\left(L_\beta + N_\beta\frac{I_{xz}}{I_{xx}}\right) + U_o\left(L_\beta\frac{I_{xz}}{I_{zz}} + N_\beta\right) - Y_r\left(L_\beta\frac{I_{xz}}{I_{zz}} + N_\beta\right) \tag{4.51}$$

$$D_2 = -Y_\beta\left(L_p N_r - L_r N_p\right) + Y_p\left(L_\beta N_r - N_\beta L_r\right) - g\cos\theta_o\left(L_\beta + N_\beta\frac{I_{xz}}{I_{xx}}\right)$$
$$+ U_o\left(L_\beta N_p - L_p N_\beta\right) - Y_r\left(L_\beta N_p - N_\beta L_p\right) \tag{4.52}$$

$$E_2 = g\cos\theta_o\left(L_\beta N_r - N_\beta L_r\right) \tag{4.53}$$

Derivations of Eqs. (4.49)–(4.53) are left to the reader as a practice problem. As observed, the lateral-directional transfer functions are functions of lateral-directional stability and control derivatives, moments of inertia, as well as the initial flight conditions [e.g., trim airspeed (U_o) and pitch angle (Θ_o)]. The applications of these transfer functions will be presented in Sect. 4.4, as well as in Chap. 6 for lateral-directional control.

4.2.3   Lateral-Directional State Space Model

Another form of dynamic formulation of the lateral-directional motion is the state space representation, which is a set of linear first-order linear differential equations, and a set of linear algebraic equations. The model is represented in the general format demonstrated in Eq. (3.42).

We are interested in state and output variables to be: (1) Side-slip angle (β), (2) Roll rate (p), (3) Yaw rate (r), and (4) Bank angle (ϕ). The yaw or heading angle (ψ) is dropped as one of the output variables, since there was a free s in the denominator of the yaw angle-to-rudder/aileron deflection. Two primary input (control) variables to lateral-directional motions are: (1) Aileron deflection (δ_A), and (2) Rudder deflection (δ_R). Hence, Eqs. (4.13)–(4.15) are reformatted in term of state variables [β, p, r, ϕ] and input variables [δ_A and δ_R] in four steps.

First, Eq. (4.13)—when $\dot{v}$ is replaced with $U_o\dot{\beta}$—is expressed in the following state space model as:

$$\dot{\beta} = \frac{1}{U_o}\left[Y_\beta\beta + Y_p p + Y_r r - U_1 r + g\phi(\cos\Theta_o) + Y_{\delta_A}\delta_A + Y_{\delta_R}\delta_R\right] \tag{4.54}$$

Second, solve Eqs. (4.14) and (4.15) to find $\dot{p}$ and $\dot{r}$. To use Cramer's rule, these two equations are reformatted into the following matrix form:

$$\begin{bmatrix} 1 & -\frac{I_{xz}}{I_{xx}} \\ -\frac{I_{xz}}{I_{zz}} & 1 \end{bmatrix}\begin{bmatrix} \dot{p} \\ \dot{r} \end{bmatrix} = \begin{bmatrix} L_\beta\beta + L_p p + L_r r + L_{\delta_A}\delta_A + L_{\delta_R}\delta_R \\ N_\beta\beta + N_p p + N_r r + N_{\delta_A}\delta_A + N_{\delta_R}\delta_R \end{bmatrix} \tag{4.55}$$

Then, $\dot{p}$ and $\dot{r}$ are determined as the ratio of two determinants:

$$\dot{p} = \frac{\left| \begin{bmatrix} L_\beta\beta + L_p p + L_r r + L_{\delta_A}\delta_A + L_{\delta_R}\delta_R & -\frac{I_{xz}}{I_{xx}} \\ N_\beta\beta + N_p p + N_r r + N_{\delta_A}\delta_A + N_{\delta_R}\delta_R & 1 \end{bmatrix} \right|}{\left| \begin{bmatrix} 1 & -\frac{I_{xz}}{I_{xx}} \\ -\frac{I_{xz}}{I_{zz}} & 1 \end{bmatrix} \right|} \tag{4.56}$$

$$\dot{r} = \frac{\left| \begin{bmatrix} 1 & L_\beta\beta + L_p p + L_r r + L_{\delta_A}\delta_A + L_{\delta_R}\delta_R \\ -\frac{I_{xz}}{I_{zz}} & N_\beta\beta + N_p p + N_r r + N_{\delta_A}\delta_A + N_{\delta_R}\delta_R \end{bmatrix} \right|}{\left| \begin{bmatrix} 1 & -\frac{I_{xz}}{I_{xx}} \\ -\frac{I_{xz}}{I_{zz}} & 1 \end{bmatrix} \right|} \tag{4.57}$$

Application of determinants yield the following:

$$\dot{p} = \frac{1}{1 - \frac{I_{xz}}{I_{zz}}\frac{I_{xz}}{I_{xx}}} \left[(L_\beta \beta + L_p p + L_r r + L_{\delta_A}\delta_A + L_{\delta_R}\delta_R) \right.$$

$$\left. + \frac{I_{xz}}{I_{xx}} (N_\beta \beta + N_p p + N_r r + N_{\delta_A}\delta_A + N_{\delta_R}\delta_R) \right] \qquad (4.58)$$

$$\dot{r} = \frac{1}{1 - \frac{I_{xz}}{I_{zz}}\frac{I_{xz}}{I_{xx}}} \left[(N_\beta \beta + N_p p + N_r r + N_{\delta_A}\delta_A + N_{\delta_R}\delta_R) \right.$$

$$\left. + \frac{I_{xz}}{I_{zz}} (L_\beta \beta + L_p p + L_r r + L_{\delta_A}\delta_A + L_{\delta_R}\delta_R) \right] \qquad (4.59)$$

These two equations seem cumbersome but can be simplified. One way is to drop the terms $\frac{I_{xz}}{I_{xx}}$ and term $\frac{I_{xz}}{I_{zz}}$ from Eqs. (4.58) and (4.59). This simplification is due to the fact that, the moment of inertia ratios $\frac{I_{xz}}{I_{xx}}$ and $\frac{I_{xz}}{I_{zz}}$ are negligible. Majority of aircraft configurations are symmetric about xz plane, so $I_{xz} \approx 0$. Thus,

$$\dot{p} = L_\beta \beta + L_p p + L_r r + L_{\delta_A}\delta_A + L_{\delta_R}\delta_R \qquad (4.60)$$

$$\dot{r} = N_\beta \beta + N_p p + N_r r + N_{\delta_A}\delta_A + N_{\delta_R}\delta_R \qquad (4.61)$$

Third, in a pure rolling motion:

$$\dot{\phi} = p \qquad (4.62)$$

Fourth and finally, Eqs. (4.54) and (4.60)–(4.62) are reformatted into matrix form as the state equation:

$$\begin{bmatrix} \dot{\beta} \\ \dot{p} \\ \dot{r} \\ \dot{\phi} \end{bmatrix} = \begin{bmatrix} \frac{Y_\beta}{U_o} & \frac{Y_p}{u_o} & -1 + \frac{Y_r}{U_o} & \frac{g\cos\Theta_o}{U_o} \\ L_\beta & L_p & L_r & 0 \\ N_\beta & N_p & N_r & 0 \\ 0 & 1 & 0 & 0 \end{bmatrix} \begin{bmatrix} \beta \\ p \\ r \\ \phi \end{bmatrix} + \begin{bmatrix} \frac{Y_{\delta_A}}{U_o} & \frac{Y_{\delta_R}}{U_o} \\ L_{\delta_A} & L_{\delta_R} \\ N_{\delta_A} & N_{\delta_R} \\ 0 & 0 \end{bmatrix} \begin{bmatrix} \delta_A \\ \delta_R \end{bmatrix} \qquad (4.63)$$

The output variables are tentatively selected to be the same as the state variables $[\beta, p, r, \phi]$. Therefore, the output equation is presented as:

$$\begin{bmatrix} \beta \\ p \\ r \\ \phi \end{bmatrix} = \begin{bmatrix} 1 & 0 & 0 & 0 \\ 0 & 1 & 0 & 0 \\ 0 & 0 & 1 & 0 \\ 0 & 0 & 0 & 1 \end{bmatrix} \begin{bmatrix} \beta \\ p \\ r \\ \phi \end{bmatrix} + \begin{bmatrix} 0 & 0 \\ 0 & 0 \\ 0 & 0 \\ 0 & 0 \end{bmatrix} \begin{bmatrix} \delta_A \\ \delta_R \end{bmatrix} \qquad (4.64)$$

In this lateral-directional state space model (Eqs. 4.63 and 4.64), A is a 4×4 square matrix, C is a 4×4 identity matrix, and B and D are 4×2 matrices, where in D, all elements are zero. The applications of this model will be presented in Sect. 4.4, as well as in Chap. 6 for lateral-directional control.

When I_{xz} is noticeable, and you would like to be more accurate, use Eqs. (4.54), (4.14), (4.15), and (4.19) as repeated here for convenience:

$$\dot{\beta} = \frac{1}{U_1}\left[Y_\beta\beta + Y_p p + r(-U_1 + Y_r) + g\phi(\cos\Theta_o) + Y_{\delta_A}\delta_A + Y_{\delta_R}\delta_R\right] \tag{4.65}$$

$$\dot{p} - \frac{I_{xz}}{I_{xx}}\dot{r} = L_\beta\beta + L_p p + L_r r + L_{\delta_A}\delta_A + L_{\delta_R}\delta_R \tag{4.66}$$

$$\dot{r} - \frac{I_{xz}}{I_{zz}}\dot{p} = N_\beta\beta + N_p p + N_r r + N_{\delta_A}\delta_A + N_{\delta_R}\delta_R \tag{4.67}$$

$$\dot{\phi} = p \tag{4.68}$$

These four equations may be expressed in the following special form of state-space model as:

$$\begin{bmatrix} 1 & 0 & 0 & 0 \\ 0 & 1 & -\frac{I_{xz}}{I_{xx}} & 0 \\ 0 & -\frac{I_{xz}}{I_{zz}} & 1 & 0 \\ 0 & 0 & 0 & 1 \end{bmatrix} \begin{bmatrix} \dot{\beta} \\ \dot{p} \\ \dot{r} \\ \dot{\phi} \end{bmatrix} = \begin{bmatrix} \frac{Y_\beta}{U_o} & \frac{Y_p}{u_o} & -1+\frac{Y_r}{U_o} & \frac{g\cos\theta_o}{U_o} \\ L_\beta & L_p & L_r & 0 \\ N_\beta & N_p & N_r & 0 \\ 0 & 1 & 0 & 0 \end{bmatrix} \begin{bmatrix} \beta \\ p \\ r \\ \phi \end{bmatrix}$$

$$+ \begin{bmatrix} \frac{Y_{\delta_A}}{U_o} & \frac{Y_{\delta_R}}{U_o} \\ L_{\delta_A} & L_{\delta_R} \\ N_{\delta_A} & N_{\delta_R} \\ 0 & 0 \end{bmatrix} \begin{bmatrix} \delta_A \\ \delta_R \end{bmatrix} \tag{4.69}$$

Since the matrix model is expressed in the following format:

$$E\dot{x} = Ax + Bu \tag{4.70}$$

The state equation will be:

$$\dot{x} = \frac{A}{E}x + \frac{B}{E}u \tag{4.71}$$

where

$$E = \begin{bmatrix} 1 & 0 & 0 & 0 \\ 0 & 1 & -\frac{I_{xz}}{I_{xx}} & 0 \\ 0 & -\frac{I_{xz}}{I_{zz}} & 1 & 0 \\ 0 & 0 & 0 & 1 \end{bmatrix} \tag{4.72}$$

The output equation is the same as the one shown in Eq. (3.64) (i.e., $y = Cx + Du$). If the aircraft asymmetry is considerable, and ratios $\frac{I_{xz}}{I_{xx}}$ and $\frac{I_{xz}}{I_{zz}}$ of an aircraft are greater than 0.05, use Eqs. (4.72), instead of (4.63).

4.2.4 Lateral-Directional Force and Moments Coefficients

The lateral-directional aerodynamic force and moments coefficients can be modelled and linearized with respect to the stability and control derivatives. These coefficients are expressed in terms of flight parameters and control surface deflections. As explained in Chap. 1, since roll and yaw rates have dimensions, they have been multiplied by $\frac{b}{2U_1}$ to make the outcome non-dimensional.

$$C_y = C_{y_\beta}\beta + C_{y_p}P\frac{b}{2U_1} + C_{y_r}R\frac{b}{2U_1} + C_{y_{\delta_a}}\delta_a + C_{y_{\delta_r}}\delta_r \tag{4.73}$$

$$C_l = C_{l_\beta}\beta + C_{l_p}P\frac{b}{2U_1} + C_{l_r}R\frac{b}{2U_1} + C_{l_{\delta_a}}\delta_a + C_{l_{\delta_r}}\delta_r \tag{4.74}$$

$$C_n = C_{n_\beta}\beta + C_{n_p}P\frac{b}{2U_1} + C_{n_r}R\frac{b}{2U_1} + C_{n_{\delta_a}}\delta_a + C_{n_{\delta_r}}\delta_r \tag{4.75}$$

The significance and calculations of the lateral-directional stability derivatives will be presented in the next few sections. Table 4.1 illustrates typical values of lateral-directional non-dimensional stability and control derivatives for fixed-wing aircraft.

Table 4.1 also demonstrates the prediction accuracy of the mathematical techniques presented to calculate the values of the lateral-directional non-dimensional stability and control derivatives.

There is a prediction accuracy in the mathematical techniques to calculate the values of the lateral-directional non-dimensional stability and control derivatives. The highest accuracy belongs to the technique to determine derivatives C_{n_β} and C_{y_p} (85%), but the lowest accuracy is for calculations of derivative C_{n_p} (only 50%). It is very hard to calculate a reliable value for this derivative.

Two better and more accurate techniques for calculations of stability and control derivatives are: (1) Wind tunnel tests, and (2) Flight tests experiments. A mathematical tool in estimation of stability derivatives using flight test experiment [1] is referred to as the maximum likelihood estimator (MLLE). Based on a modified version of this technique, Ref. [2] presents methods and practical considerations for stability derivatives for conventional transonic aircraft.

Table. 4.1 Lateral-directional non-dimensional stability and control derivatives

Derivative	Typical values (1/rad)		Prediction accuracy (%)
	Minimum	Maximum	
C_{l_β}	-0.4	$+0.1$	80
C_{y_β}	-0.1	-2	80
C_{n_β}	0.05	0.4	85
C_{l_p}	-0.1	-0.8	80
C_{y_p}	-0.3	$+0.8$	85
C_{n_p}	-0.5	$+0.1$	50
C_{l_r}	0	$+0.6$	70
C_{y_r}	0	1.2	60
C_{n_r}	0	-1	75
$C_{l_{\delta_A}}$	0.05	0.4	85
$C_{l_{\delta_R}}$	-0.05	0.04	80
$C_{n_{\delta_A}}$	-0.08	0.08	70
$C_{n_{\delta_R}}$	0	-0.15	85
$C_{y_{\delta_R}}$	0	0.5	80

4.3   Directional Static Stability

Directional stability refers to the tendency of an aircraft to oppose any directional disturbance (e.g., a gust to the left fuselage nose), and to return to its present directional state of flying motion, if disturbed in the xy plane. Directional static stability is defined as to the tendency of an aircraft to develop forces and moments to oppose any directional disturbance/perturbations (e.g., yawing disturbance).

In a fixed-wing air vehicle, the directional stability is mainly provided by the vertical tail (or stabilizer) with a desired yawing moment arm. To have directional static stability, aircraft needs to develop a restoring (i.e., negative) yawing moment (N), when it is displaced from its directional equilibrium point (often $\beta = 0$).

The most important flight parameter in the directional plane is the sideslip angle (β). In a cruising flight, the sideslip angle is usually zero. Moreover, in a turning flight, a small sideslip angle is generated to keep the turn coordinated. The zero sideslip angle in a cruising level flight, and a small value for sideslip angle in a turning flight, is part of the directional equilibrium/trim condition.

Consider a gust hit at the right of fuselage nose, or a gust hit at to the left of the vertical tail and then quickly disappears. This directional disturbance rotates the aircraft about z-axis and is generating a negative sideslip angle (i.e., nose to the right, $-\Delta\beta$). In a statically directionally stable aircraft, a positive yawing moment ($+\Delta$ N) is required (see

Fig. 4.2) to oppose the disturbance and restore the trim point (i.e., $\beta = 0$). Similarly, when a positive sideslip angle $(+\Delta\beta)$ is generated by a directional disturbance, a positive (i.e., nose to the left) yawing moment $(+\Delta N)$ is desired to oppose the disturbance and restore the directional trim point (i.e., $\beta = 0$). The aerodynamic yawing moment is a function of airspeed (V), air density (ρ), geometry (S and b), and configuration:

$$N = \frac{1}{2}\rho V^2 SbC_n \tag{4.76}$$

For static directional stability analysis, the yawing moment coefficient (C_n) is a good representative of aircraft configuration. This static stability requirement can be visualized as the rate of change of yawing moment coefficient due to a change in sideslip angle (β), which is modelled as a new parameter, $C_{n\beta}$. Thus, the requirement for **directional static stability** is just to have a positive value for this parameter.

$$C_{n\beta} = \frac{\partial C_n}{\partial \beta} > 0 \tag{4.77}$$

The higher the value, the aircraft will be more directionally statically stable. $C_{n\beta}$— also referred to as *yaw stiffness derivative*—is the variation of aircraft yawing moment coefficient with dimensionless rate of change of angle of sideslip.

Unlike the case of pitch, the yawing moment coefficient versus yaw angle must have a positive slope for stability. The reasoning is identical to that previously used in the pitch case; the vertical fin must be able to create a restoring moment that minimizes the yaw angle caused by a side force disturbance. A typical variations of yawing moment coefficient with respect to the sideslip angle is shown in Fig. 4.3. In some references, this stability is referred to as weathercock stability, since an aircraft possessing static directional stability will always point into the relative wind.

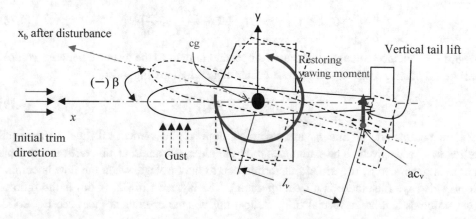

Fig. 4.2 Directional disturbance and the opposing yawing moment (Top view)

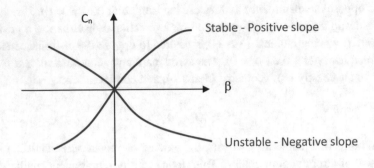

Fig. 4.3 Static directional stability

The most important flight parameters in the directional plane (xy) are: (1) Sideslip angle (β), (2) Heading angle (ψ) and (3) Yaw rate (r). In a statically directionally stable aircraft, only disturbance to sideslip angle and yaw rate are opposed. Note that, the directional stability is not the same as the heading stability, since no aircraft features inherent heading stability. An aircraft does not have any memory of its heading angle, which is defined as the difference between current flight direction and a reference line (e.g., North direction). If a wind gust changes the heading of an aircraft, the aircraft does not have a tendency to return to its initial direction. To keep an aircraft on a desired heading angle, an automatic flight control system is required.

The C_{n_β} is a non-dimensional stability derivative, and is referred to as the directional static stability derivative (also yaw-stiffness derivative). For an aircraft with a fixed vertical tail, the aircraft static directional stability derivative is determined by adding wing, fuselage, and vertical tail contributions. The wing contribution at low angle of attack tends to be negligible. For fuselage contribution, see references such as [3]. The vertical tail contribution is determined as:

$$C_{n_\beta} \cong C_{n_{\beta_v}} = C_{L_{\alpha v}}\left(1 - \frac{d\sigma}{d\beta}\right)\eta_v \frac{S_v l_v}{Sb} \tag{4.78}$$

where $C_{L_{\alpha v}}$ is the vertical tail lift curve slope, and l_v refer to the distance between vertical tail aerodynamic center to the aircraft center of gravity (see Fig. 4.2).

$$l_v = x_{ac_v} - x_{cg} \tag{4.79}$$

The parameters $\frac{d\sigma}{d\beta}$ and η_v are sidewash slope at the vertical tail, and vertical tail efficiency respectively. The parameter σ is the sidewash angle at the vertical tail. This angle is generated by the airfield distortion due to the fuselage, when the flow is coming from sideways (i.e., there is a sideslip angle). The last term in Eq. (4.66) is the ratio of two volumes and has a special title, vertical tail volume coefficient ($\overline{V}_V$, see Eq. 2.43).

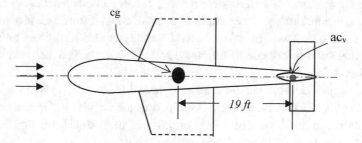

Fig. 4.4 General aviation aircraft for Example 4.1 (not to scale)

The value for C_{n_β} of a directionally statically stable aircraft ranges from $+ 0.1$ to $+ 0.4$ 1/rad.

Example 4.1 Consider a fixed-wing General Aviation aircraft—shown below—with a maximum takeoff weight of 2500 lb, a wing planform area of 164 ft^2, and the following characteristics:

$$\frac{d\sigma}{d\beta} = 0.21; \quad C_{L_{\alpha v}} = 4.3 \, \frac{1}{rad}; \quad b = 32 \, ft; \quad \eta_v = 0.98; \quad S_v = 36 \, ft^2$$

Determine directional static statically derivative; C_{n_β}. Is this aircraft directionally statically stable?

Solution

From Fig. 4.4, the vertical tail ac is 19 ft from aircraft center of gravity ($l_v = 19$ ft). The longitudinally static statically derivative:

$$C_{n_\beta} = C_{L_{\alpha v}} \left(1 - \frac{d\sigma}{d\beta} \right) \eta_v \frac{S_v l_v}{Sb}$$

$$C_{n_\beta} = 4.3 \times (1 - 0.21) \times 0.98 \times \frac{36 \times 19}{175 \times 32} = 0.43 \, \frac{1}{rad} \qquad (4.78)$$

The aircraft is directionally statically stable, since the C_{n_β} is positive.

4.4 Lateral Static Stability

Lateral stability is the inherent tendency of an aircraft to oppose any lateral disturbance, and to return to its present lateral trim condition, if disturbed. Lateral static stability is defined as to the tendency of an aircraft to develop forces and moments to oppose any lateral disturbance/perturbations. An example of a usual lateral disturbance is a vertical gust when under a wing tip (see Fig. 4.1b). Lateral stability is also significant as directional and longitudinal stability because an unwanted roll will cause a bank and may finally end up a dangerous spin.

Two most important flight parameters in the lateral plane are: (1) Sideslip angle (β), and (2) the bank angle (ϕ). Bank angle is defined as the angular difference between the aircraft xy plane and horizontal. In a level cruising flight, both sideslip and bank angles are zero. However, in a coordinated turning flight, a desired bank angle and a Sideslip angle are desired to coordinate the turn. The zero bank/sideslip angles in a cruise, or desired values for bank/sideslip angles in a coordinated turn are part of the lateral trim condition. Lateral static stability is the opposition of an aircraft to any change in the current trim values of the sideslip and bank angles.

When the aircraft bank angle is disturbed in a level flight, the vehicle begins to roll, which trigs two other undesired outputs: (1) Loss of altitude, (2) Sideslip angle. To have lateral static stability, aircraft needs to develop a restoring rolling moment (L), when it is displaced from its lateral equilibrium point. Note that, the restoring rolling moment will not resolve the loss of altitude, but it will first restore the initial zero side slip angle (β). Moreover, along with restoring the sideslip angle, the bank angle is also restored.

If a positive sideslip angle ($+ \Delta\beta$) is generated by a lateral disturbance (see Fig. 4.5), a positive rolling moment ($+ \Delta L$) is desired to oppose the disturbance. Similarly, when a negative sideslip angle ($- \Delta\beta$) is generated by a positive rolling disturbance, a negative ($-\Delta L$) rolling moment is desired to oppose the disturbance.

The entire process may be divided into the following five steps: (a) aircraft is laterally trimmed, and wing is level, no sideslip angle; (b) a vertical gust hits under the left wing; (c) when gust rolls the aircraft (positively; left wing up and right wing down) disappears, a positive sideslip angle is experienced; (d) A negative rolling moment is generated to oppose this disturbance; and e. aircraft returns back to wing level, and zero-sideslip angle.

When an aircraft is disturbed from a wing-level attitude, it will begin to sideslip as shown in Fig. 4.4c. Note that, the reason for a positive sideslip angle is the following. When the aircraft banks (here, to the right), the vertical component of the lift will be less

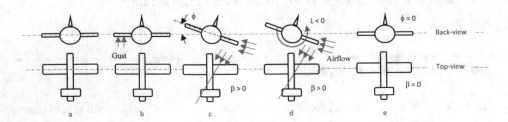

Fig. 4.5 Lateral disturbance and the opposing rolling moment (top and back views)

than the aircraft weight. This causes the aircraft to lose altitude, while moving forward and also to the right simultaneously. So, the airflow (relative wind) will come from the front plus right side. A positive sideslip angle is when the airflow is coming from the right-side (see Fig. 1.4).

The aerodynamic rolling moment is a function of airspeed (V), air density (ρ), geometry (S and b), and configuration:

$$L = \frac{1}{2}\rho V^2 SbC_l \tag{4.80}$$

For static lateral stability analysis, the rolling moment coefficient (C_l) is a good representative of aircraft configuration. This static stability requirement can be visualized as the rate of change of rolling moment coefficient due to a change in sideslip angle (β), which is modelled as a new parameter, C_{l_β}. Thus, the requirement for lateral static stability (see Fig. 4.6) is just to have a negative value for this parameter.

$$C_{l_\beta} = \frac{\partial C_l}{\partial \beta} < 0 \tag{4.81}$$

The higher the negative value, the aircraft will be more laterally statically stable. This non-dimensional stability derivative is often referred to as the aircraft "*dihedral effect*" or "lateral static stability derivative", or "*dihedral derivative*".

For a fixed-wing aircraft, the static lateral stability derivative is determined by adding wing, fuselage, horizontal tail, and vertical tail contributions:

$$C_{l_\beta} = C_{l_{\beta_{wf}}} + C_{l_{\beta_h}} + C_{l_{\beta_v}} \tag{4.82}$$

Methods for estimating the aircraft components contribution to C_{l_β} can be found in [4]. The major contributor to C_{l_β} is the wing dihedral angle (Γ). The wing dihedral angle is the angle that the wing sections make with the fuselage as you view from front view

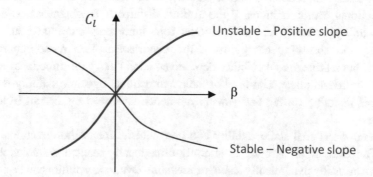

Fig. 4.6 Static lateral stability

(i.e., in the yz plane). Typical value for C_{l_β} of a laterally statically stable aircraft can range from -0.4 to $+0.1$ 1/rad, but may change significantly with Mach number in the transonic speed.

Note that, the lateral stability is not the same as the rolling stability, and aircraft does not often feature a roll stability. An aircraft does not often have any memory of its initial bank angle, but it has the memory of its sideslip angle. If a gust changes only the bank angle, and not the sideslip angle (e.g., in a pure rolling motion), the aircraft does not have a tendency to return to its bank angle. To keep an aircraft on a desired bank angle, an automatic flight control system is required. Hence, no aircraft have inherent roll stability, but majority of fixed-wing aircraft have inherent lateral stability.

Due to this fact, and a compromise in design, majority of aircraft are lightly statically laterally stable. It implies that the aircraft reaction to roll upset is not often fast enough as desired. Thus, majority of aircraft are equipped with a roll damper or a wing leveler to artificially stabilize the aircraft in lateral mode. It is interesting to know that the first autopilot in history was developed in 1912 by Sperry Corporation to keep the wing level. This is a very beneficial flight aid to a human pilot in a long cruising flight, so that fatigue is reduced by allowing the pilot to have his/her hands off the control stick.

In comparing lateral static stability with directional static stability, almost all fixed-wing aircraft with a conventional configuration have directional static stability. However, many fixed-wing aircraft with a conventional configuration are not laterally statically stable.

4.5   Lateral-Directional Dynamic Stability

4.5.1   Fundamentals

Fundamentals, definition, and levels of dynamic stability is discussed in Chap. 3, Sect. 3.4.1. Lateral-directional dynamic stability is the tendency to return to the present lateral-directional trim conditions, if the aircraft disturbed. To obtain lateral-directional dynamic stability, restoring lateral-directional forces/moments must have the capability of damping and absorbing energy from the disturbance. Two most important non-dimensional lateral-directional stability derivatives are: (1) C_{n_r} for directional plane, and (2) C_{l_p} for the lateral plane. The level of lateral-directional dynamic stability is based on the power of restoring rolling and yawing moments developed by aircraft in to damp a disturbance.

The lateral-directional static stability is a pre-requisite for lateral-directional dynamic stability. However, if an aircraft is laterally-directionally statically stable, it may be laterally-directionally dynamically stable or unstable. Dynamic stability can be augmented by artificial means, such as an autopilot using feedback, which are used to control the air vehicle. A very popular stability augmenter for a large transport aircraft is the yaw damper which augments the directional stability.

Dynamic stability can be augmented by artificial means, such as an autopilot which primarily is used to control the air vehicle. A very popular stability augmenter for large transport aircraft is the yaw damper which augments the directional dynamic stability. However, dynamic instability may arise, if the controller and feedback in the control system are improperly designed. This can lead to an amplification of the instability instead of damping.

4.5.2 Dynamic Stability Analysis Technique

Analysis of lateral-directional dynamic stability is much more complex than that for the lateral-directional static stability, and requires a lot of calculations. The analysis may be conducted in either of two domains: (1) Frequency domain, (2) Time domain.

In frequency domain analysis, transfer functions (in s-domain) are utilized. In practice, only denominator of transfer functions suffices. As derived in Sect. 4.3, the denominators of all lateral-directional transfer functions are similar and is a polynomial as a function of s. When this equation is equated to zero, we obtain the following:

$$s\left[A_2 s^4 + B_2 s^3 + C_2 s^2 + D_2 s + E_2\right] = 0 \tag{4.83}$$

This is a fifth-order polynomial in s-domain, with a free s. The coefficients A_2, B_2, C_2, D_2, and E_2 are real quantities and have been already presented. Since the lateral-directional absolute stability can be analyzed directly from these coefficients, this equation is referred to as the lateral-directional characteristic equation. The order, signs, coefficients, and roots of this equation reveal a lot of information about characteristics of the aircraft with regards to lateral-directional dynamic stability.

In time domain analysis, either the governing differential equations of motion, or state space model of dynamics are employed. In practice, only matrix A (i.e., state matrix) for lateral-directional motion ($A_{lat\text{-}dir}$) suffices:

$$A_{lat-dir} = \begin{bmatrix} \frac{Y_\beta}{U_o} & \frac{Y_p}{u_o} & -1 + \frac{Y_r}{U_o} & \frac{g\cos\Theta_o}{U_o} \\ L_\beta & L_p & L_r & 0 \\ N_\beta & N_p & N_r & 0 \\ 0 & 1 & 0 & 0 \end{bmatrix} \tag{4.84}$$

Elements of matrix $A_{lat\text{-}dir}$ have been presented earlier in this chapter.

In the lateral-directional dynamic stability analysis in s-domain, the following four steps are conducted:

1. Calculate values for all aircraft lateral-directional stability derivatives;
2. Write lateral-directional characteristic equation (Eq. 3.83);
3. Calculate roots of this equation (five roots);
4. If roots have all negative real parts, the air vehicle is said to be dynamically laterally-directionally stable. If there is at least one root with a positive real part, the aircraft

is dynamically laterally-directionally unstable. However, if at least real part of one root is positive, and all other roots have negative real parts, the aircraft is dynamically laterally-directionally neutrally stable.

Note: The lateral-directional characteristic equation is a fifth-order polynomial in s-domain, with a free s (i.e., s = 0). However, in all lateral-directional transfer functions except the heading-angle (to aileron and rudder) transfer functions, there is a free s in the numerator. When, both two "s" are cancelled from numerator and denominator, non- heading-angle transfer functions have no free s and have fourth-order characteristic equation. The free s in the denominator of the heading-angle-to-aileron/rudder-deflection transfer functions is an indication of neutral stability with respect to changes in heading angle. Hence, in analyzing in s-domain, we only consider the fourth order polynomial; inside parenthesis:

$$A_2 s^4 + B_2 s^3 + C_2 s^2 + D_2 s + E_2 = 0 \qquad (4.85)$$

In the lateral-directional dynamic stability analysis in time-domain, the following four steps are conducted:

1. Calculate values for all aircraft lateral-directional dimensional stability derivatives;
2. Write state matrix (i.e., A) of lateral-directional state space model (Eq. 3.84);
3. Calculate eigenvalues of matrix A (four eigenvalues);
4. If eigenvalues have all negative real parts, the air vehicle is said to be dynamically laterally-directionally stable. If there is at least one eigenvalue with a positive real part, the aircraft is dynamically laterally-directionally unstable. However, if at least real part of one eigenvalue is positive, and all other eigenvalues have negative real parts, the aircraft is dynamically laterally-directionally neutrally stable.

Example 4.2 A twin-engine GA aircraft has the following sideslip-angle-to-rudder-deflection transfer function for a cruising flight:

$$\frac{\beta(s)}{\delta_R(s)} = \frac{835 s^2 + 380 s + 270}{181 s^4 + 392 s^3 + 820 s^2 + 1220 s - 25}$$

Is the aircraft dynamically laterally-directionally stable? Why?

Solution
The lateral-directional characteristic equation (i.e., denominator) is:

$$181 s^4 + 392 s^3 + 820 s^2 + 1220 s - 25 = 0$$

Four roots of lateral-directional characteristic equation are: $-0.205 \pm 1.95i$; -1.78; $+0.0202$. Two roots are complex (two complex pairs), and two roots are real. Since there is a positive real root, the aircraft is dynamically laterally-directionally unstable. Hence, there is a lateral-directional dynamic instability.

Example 4.3 A large transport aircraft has the following lateral-directional state space model for a cruising flight at $M = 0.9$ at 35,000 ft:

$$
\begin{bmatrix} \dot{\beta} \\ \dot{p} \\ \dot{r} \\ \dot{\phi} \end{bmatrix} = \begin{bmatrix} 0 & 0.02 & -1 & -0.3 \\ -2.2 & 0 & 0.03 & -17 \\ 0.04 & 0 & -0.33 & 8.1 \\ 0 & 1 & 0 & 0 \end{bmatrix} \begin{bmatrix} \beta \\ p \\ r \\ \phi \end{bmatrix} + \begin{bmatrix} 0 & 0.02 \\ 5 & 4.3 \\ -0.2 & -3.1 \\ 0 & 0 \end{bmatrix} \begin{bmatrix} \delta_A \\ \delta_R \end{bmatrix}
$$

$$
\begin{bmatrix} \beta \\ p \\ r \\ \phi \end{bmatrix} = \begin{bmatrix} 1 & 0 & 0 & 0 \\ 0 & 1 & 0 & 0 \\ 0 & 0 & 1 & 0 \\ 0 & 0 & 0 & 1 \end{bmatrix} \begin{bmatrix} \beta \\ p \\ r \\ \phi \end{bmatrix} + \begin{bmatrix} 0 & 0 \\ 0 & 0 \\ 0 & 0 \\ 0 & 0 \end{bmatrix} \begin{bmatrix} \delta_A \\ \delta_R \end{bmatrix}
$$

Is the aircraft dynamically laterally-directionally stable? Why?

Solution

The state matrix (A) of the lateral-directional state space model is:

$$
A = \begin{bmatrix} 0 & 0.02 & -1 & -0.3 \\ -2.2 & 0 & 0.03 & -17 \\ 0.04 & 0 & -0.33 & 8.1 \\ 0 & 1 & 0 & 0 \end{bmatrix}
$$

Four eigenvalues of this matrix are: $-0.032 \pm 4.24i$; -1.12; $+0.85$. Two roots are complex (two complex pairs), and two roots are real. Since there is a positive real root, the aircraft is dynamically laterally-directionally unstable. Hence, there is a lateral-directional dynamic instability.

4.5.3 Lateral-Directional Response Modes

As discussed in Chap. 2, lateral and directional motions in a fixed-wing aircraft is coupled due to vertical tail and sideslip. The interaction between the roll and the yaw impacts the aircraft response to a lateral-directional disturbance.

When a dynamically laterally-directionally stable aircraft responds to a lateral-directional disturbance, lateral-directional flight variables (e.g., sideslip angle, bank angle, yaw angle, roll rate, and yaw rate) will often follow either a damped/undamped oscillatory (often a sinusoidally) motion, or an exponentially damped/undamped motion. Moreover, some of the flight parameters are highly damped, while others are lightly damped.

Any proper fourth order transfer function -with two complex conjugates roots, and two real roots—can be reformatted as the combination of one second order and two first-order transfer functions:

$$\frac{Num}{A_2 s^4 + B_2 s^3 + C_2 s^2 + D_2 s + E_2} = \frac{Num_1}{s^2 + a_{dr}s + b_{dr}} + \frac{Num_2}{s + a_s} + \frac{Num_3}{s + a_r} \quad (4.86)$$

where Num stands for numerator. Recall that a second-order system (see Fig. 4.7a) is characterized by a natural frequency (or period) and a damping ratio, while a first order stable system (see Fig. 4.7a) is featured by a time constant.

The response of an aircraft to a lateral-directional disturbance is a complicated combination of rolling, yawing, and sideslipping motions. This reaction may be categorized into three modes: (1) Dutch roll, (2) Spiral, and (3) Roll subsidence mode. If all these response modes are damped, the air vehicle is laterally-directionally dynamically stable.

The second order oscillatory mode ($s^2 + a_{dr}s + b_{dr}$) has an oscillatory motion that resembles to the motion of a drunken skater, so it is referred to as the *dutch roll* mode (see Fig. 4.8). This mode is a complex combination of rolling, yawing, and side slipping and often has a low frequency. This mode is represented by a pair of complex roots or eigenvalues.

Roll subsidence (or just roll) mode is also a first order non-oscillatory (i.e., exponentially) mode that involves mostly roll rate and a corresponding bank angle with time (see Fig. 4.9a). This mode is represented by a real root or eigenvalue ($s = -a_r$). It often has a shorter time-constant (about one second), which indicates a fast roll response.

Spiral is a first order non-oscillatory (i.e., exponentially) mode that is the change in yaw (heading) angle with time (see Fig. 4.9b). This mode is represented by a real root or eigenvalue ($s + a_s$). It often has a longer time-constant (about one minute), which indicates a slow yaw response. In some aircraft, the spiral mode may be unstable, which cause an aircraft to get into an ever-steeper spiral dive.

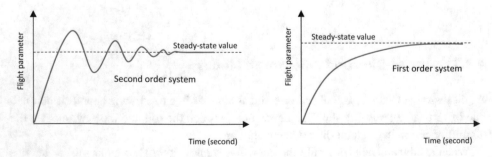

Fig. 4.7 General form of first and second order stable systems

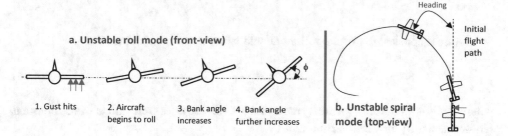

Fig. 4.8 Unstable spiral and unstable roll modes

Fig. 4.9 Unstable dutch roll
mode (front-view)

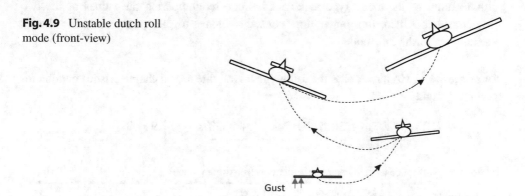

In majority of fixed-wing aircraft with a conventional configuration, dutch roll mode and roll mode are highly damped, but spiral mode is either slowly damped (i.e., convergent) or undamped (i.e., divergent motion). When dutch roll mode is lightly damped, the landing in gusty wind conditions would be difficult for the pilot. If the damping of this mode is not artificially augmented, it may upset pilots and passengers. Moreover, in a transport aircraft, passengers sitting in rear fuselage seats would be very uncomfortable in turbulent conditions. The period can be on the order of 5–10 s.

The damping ratio (ζ) and undamped natural frequency (ω_n) of dutch roll mode are obtained by simultaneously solving the following two equations:

$$\eta = -\zeta \omega_n \tag{4.87}$$

$$\omega = \omega_n \sqrt{1 - \zeta^2} \tag{4.88}$$

where ω is the imaginary part and η is the real part of the complex roots ($s_{1,2} = \eta \pm i\omega$). The period is:

$$t_{dr} = \frac{2\pi}{\omega_n}$$ (4.89)

The time constant (τ) of the roll mode and spiral mode are obtained from:

$$\tau = \frac{1}{-a}$$ (4.90)

where "a" is the value of real root. Table 4.2 provides roll mode time constant requirements [5] for Classes I through IV.

The requirements of the damping ratio and frequency of the dutch roll are specified in Table 4.3. The frequency and damping ratio must exceed the values given in this table.

Allocation of the mode type to a root is done by comparing the values of the two time-constants. Often, the longer time constant belongs to the spiral, while the shorter one belongs to the roll mode.

Example 4.4 A GA aircraft has the following lateral-directional characteristic equation for a cruising flight:

$$207s^4 + 3045s^3 + 7426s^2 + 38587s - 337.9 = 0$$

Table. 4.2 Roll mode time constant specification (maximum value)

Flight phase	Aircraft class	T_R (s)		
		Level 1	Level 2	Level 3
A	I, IV	1.0	1.4	10
	II, III	1.4	3.0	10
B	All	1.4	3.0	10
C	I, IV	1.0	1.4	10
	II, III	1.4	3.0	10

Table. 4.3 Dutch roll mode handling qualities

Level	Flight phase	Aircraft class	min ζ_d	min $\zeta_d\omega_{n_d}$ (rad/s)	min ω_{n_d} (rad/s)
1	A	I, IV	0.19	0.35	1.0
		II, III	0.19	0.35	0.4
	B	All	0.08	0.15	0.4
	C	I, II, IV	0.08	0.15	1.0
		III	0.08	0.15	0.4
2	All	All	0.02	0.05	0.4
3	All	All	0.02	No limit	0.4

Identify the dutch roll, roll, and spiral modes of lateral-directional dynamics; and then determine damping ratio, natural frequency, and period of the dutch roll mode, and time constants of roll mode and spiral mode. Is the aircraft dynamically lateral-directionally stable? Why?

Solution

Four roots of lateral-directional characteristic equation are: $-0.83\pm3.69i$; -13.02; $+0.009$. Two roots are complex (a complex pairs), and two roots are real. Since one root has a positive real part, the aircraft is dynamically laterally-directionally unstable.

However, we can identify which mode is unstable. The complex conjugate often represents the dutch roll mode. Since this pair has a negative real part, the dutch roll mode is stable and will be damped. The real part with a negative value often represents a damped roll mode, and the real part with a positive value often represents an undamped spiral mode. Thus,

– Dutch roll: $\frac{Num}{s^2+1.66s+14.3} = \frac{Num}{s^2+2\zeta\omega_n s+\omega_n^2}$ (second order oscillatory mode): Damped
– Roll: $\frac{Num}{s+13.02}$ (first order mode): Damped
– Spiral: $\frac{Num}{s-0.009}$ (first order mode): Un-damped

a. Dutch roll mode.

Undamped natural frequency, damping ratio, and period:

$$s^2 + 1.66s + 14.3 = s^2 + 2\zeta\omega_n s + \omega_n^2 \Rightarrow$$

$$\omega_n = \sqrt{14.3} = 3.78 \frac{rad}{s}$$

$$2\zeta\omega_n = 1.66 \Rightarrow \zeta = \frac{1.66}{2 \times 3.78} \Rightarrow \zeta = 0.22$$

$$t_{dr} = \frac{2\pi}{\omega_{dr}} = \frac{2\pi}{\omega_n\sqrt{1-\zeta^2}} = \frac{2\pi}{3.78\sqrt{1-0.22^2}} = 1.704 \, s$$

b. Roll time constant: $\tau_s = \frac{1}{s_2} = \frac{1}{13.02} = 0.077 \, s$
c. Spiral time constant: Since un-damped, there is no time constant.

The aircraft is laterally-directionally dynamically unstable, since there is one positive root.

The characteristics of dutch roll, spiral, and roll subsidence modes for laterally-directionally stable aircraft can be predicted from an approximation to the lateral-directional equations of motion. In this approximation, natural frequency ($\omega_{n_{dr}}$) and damping ratio (ζ_{dr}) of dutch roll mode are expressed as functions of dimensional stability derivatives:

$$\omega_{n_{dr}} = \sqrt{\frac{Y_\beta N_r - N_\beta Y_r + U_1 N_\beta}{U_1}} \qquad (4.91)$$

$$\zeta_{dr} = \frac{-1}{2\omega_{n_{dr}}}\left(\frac{Y_\beta + U_1 N_r}{U_1}\right) \qquad (4.92)$$

These equations are derived, when the Dutch roll mode is considered to consist primarily of sideslipping and yawing motions.

The time constant (τ) of the roll mode is estimated by considering a single degree of freedom pure rolling motion:

$$\tau_r = -\frac{1}{L_p} \qquad (4.93)$$

The time constant (τ) of the spiral mode is estimated by neglecting the side force equation and considering rolling and yawing moments equations:

$$\tau_s = -\frac{L_\beta}{L_\beta N_r - N_\beta L_r} \qquad (4.94)$$

These equations provide a reasonable and good approximation (about 90% accuracy) for subsonic speeds with low angles of attack.

4.6   Aircraft Reaction to Lateral-Directional Disturbances

In order for an aircraft to feature dynamic lateral-directional stability, restoring side force, and rolling and yawing moments must be developed such that the flight motion disturbance is damped. Dynamic lateral-directional stability is created by aerodynamic forces and moments that are proportional to the rate and reaction of the various surfaces such as wing and vertical tail. Each aircraft configuration has a unique contribution to the overall response to any disturbance. In this section, the aircraft responses to various longitudinal disturbance are discussed.

The perturbed-state forces and moments—created by disturbance—are temporary and additional to steady-state values of forces and moments. The perturbed-state forces and moments are temporarily generated to react to a disturbance, while the steady-state forces and moments are essential to provide the aircraft equilibrium. When the aircraft returns to lateral-directional trim condition, the perturbed-state forces and moments disappear.

Dynamic lateral-directional stability can be translated into aircraft disturbance damping power, which is mathematically modelled as the lateral-directional stability derivatives. There are mainly five direct longitudinal disturbance: (1) Disturbance to bank angle $(\Delta\phi)$, (2) Disturbance to sideslip angle $(\Delta\beta)$, (3) Disturbance to yaw angle $(\Delta\psi)$, (4) Disturbance to roll rate (Δp), (5) Disturbance to yaw rate (Δr).

Primary lateral-directional stability derivatives are C_{y_β}; C_{y_p}; C_{y_r}; C_{l_β}; C_{l_p}; C_{l_r}; C_{n_β}; C_{n_p}; C_{n_r}, which less or more represent aircraft reaction to lateral-directional disturbances. The importance and application of these derivatives will be addressed in the following section.

4.6.1 Aircraft Reaction to Disturbance in Bank Angle

In a symmetric aircraft, a disturbance in the bank angle—if does not disturb the sideslip angle—will not generate any aircraft reaction in terms of side force, rolling moment and yawing moment. Thus, an aircraft does not often feature a rolling stability, and does not often have any memory of its bank angle. If a gust changes only the bank angle, and not the sideslip angle (e.g., in a pure rolling motion), the aircraft does not have a tendency to return to its initial trim bank angle.

However, in a real flight, when a gust creates a non-zero bank angle; a sideslip will be generated (i.e., non-pure rolling motion). Consider an aircraft in a level flight (where $\phi = 0$), while the bank angle is disturbed (i.e., moving away from level flight). This disturbance causes one wing section to drop, while the other section rises. Now, the vertical component of the lift is less than the aircraft weight, which means the aircraft will begin to descend, while slip to one side (toward the lowered wing section).

The relative wind due to sideslip will affect the entire aircraft and will be from the direction of the slip. When a disturbance in the bank angle changes the trim sideslip angle, it causes the aircraft to react in terms of side force, rolling moment, and yawing moment. These reactions will be discussed in the following section.

4.6.2 Aircraft Reaction to Disturbance in Sideslip Angle

A disturbance sideslip angle may be generated by two ways: (1) Directly when a horizontal gust hits the aircraft from one side, (2) After a roll upset (due to a vertical gust under one side of the wing), the aircraft descends and sideslips due to the bank angle (Fig. 4.10).

The disturbance in sideslip will cause three separate and parallel aircraft reactions: (1) Aerodynamic rolling moment, (2) Aerodynamic yawing moment, (3) Aerodynamic side force. These perturbed-state moments and force are discussed here. A sideslip angle perturbation will induce a local change in angle of attack on vertical tail.

a. **Aerodynamic rolling moment due to a sideslip angle disturbance**

When an aircraft is cruising with a level flight, a disturbance in sideslip angle (β) will cause a perturbed-state rolling moment (L). This reaction is modeled as the dimensional

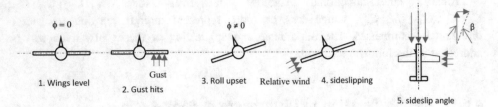

Fig. 4.10 A disturbance in bank angle creates a sideslip angle (front-view, top-view)

derivative L_β which is obtained by:

$$L_\beta = \frac{\overline{q}_1 S b C_{l_\beta}}{I_{xx}} \tag{4.95}$$

The non-dimensional derivative C_{l_β} is the rolling moment coefficient due to sideslip derivative ($\frac{\partial C_l}{\partial \beta}$) and is determined by adding the contribution of win-fuselage, horizontal tail, and vertical tail (Eq. 4.82). This non-dimensional stability derivative—by some references—is referred to as the dihedral effect. The main contribution to this derivative is coming from wing via its three geometric parameters: (1) Wing dihedral angle, (2) Wing position on the fuselage (e.g., high, low), and (3) Sweep angle.

The process of calculation of C_{l_β} is very long; the interested reader may refer to [4] or part VI of Ref. [6] for a detailed estimation method. The derivative C_{l_β} should be negative to have a positive contribution in static/dynamic lateral-directional stability.

b. **Aerodynamic yawing moment due to a sideslip angle disturbance**

A disturbance in sideslip angle (β) to an aircraft—when it is cruising with a level flight—will cause a perturbed-state yawing moment (N). This reaction is modeled as the dimensional derivative N_β which is obtained by:

$$N_\beta = \frac{\overline{q}_1 S b C_{n_\beta}}{I_{zz}} \tag{4.96}$$

The non-dimensional derivative C_{n_β} is the yawing moment coefficient due to sideslip derivative ($\frac{\partial C_n}{\partial \beta}$) and is determined by adding the contribution of win-fuselage, horizontal tail, and vertical tail (Eq. 4.78). The derivative C_{n_β} should be positive to have a positive contribution in static/dynamic lateral-directional stability.

Example 4.5 A jet airliner with the following characteristics is cruising with a speed of 200 knot at sea level.

$$S = 20\,\text{m}^2;\ S_v = 3\,\text{m}^2;\ b = 12\,\text{m};\ l_v = 10\ \text{m};$$

$$d\sigma/d\beta = 0.1; C_{L_{\alpha_v}} = 4.7\frac{1}{\text{rad}}; \eta_v = 0.96$$

A 30 knot wind gust is hitting from right side, and imposes a sideslip angle. What yawing moment is generated by the aircraft as an inherent response to this disturbance?

Solution

$$\beta = \tan^{-1}\left(\frac{v}{U_o}\right) = \tan^{-1}\left(\frac{30}{200}\right) = 0.149 \text{ rad} = 8.53 \text{ deg} \tag{1.3}$$

$$C_{n_\beta} = C_{L_{\alpha_v}}\left(1 - \frac{d\sigma}{d\beta}\right)\eta_v\frac{S_v l_v}{Sb} = 4.7 \times (1 - 0.1) \times 0.96\frac{3 \times 10}{20 \times 12} = 0.529\frac{1}{\text{rad}} \tag{4.78}$$

This problem can be solved by using either: (1) dimensional derivative N_β (Eq. 4.18) or (2) yawing moment equation (Eq. 4.72) directly. Both will lead to the same result:

$$C_n = C_{n_\beta}\beta = 0.529 \times 0.149 = 0.079 \tag{4.75}$$

$$N = \frac{1}{2}\rho V^2 SbC_n = \frac{1}{2} \times 1.225 \times (200 \times 0.514)^2$$
$$\times 20 \times 12 \times 0.079 = 122,500 \text{ Nm} \tag{4.76}$$

c. **Aerodynamic side force due to a sideslip angle disturbance**

A disturbance in sideslip angle (β) to an aircraft—when it is cruising with a level flight—will cause a perturbed-state side force (Y). This reaction is modeled as the dimensional derivative Y_β which is obtained by:

$$Y_\beta = \frac{-\bar{q}_1 S C_{y_\beta}}{m} \tag{4.97}$$

The non-dimensional derivative C_{y_β} is the side force coefficient due to sideslip derivative ($\frac{\partial C_y}{\partial \beta}$) and is determined by adding the contribution of win-fuselage, horizontal tail, and vertical tail. However, the wing contribution with a low dihedral angle tends to be negligible. The fuselage contribution varies and depends on the shape and size of the fuselage. The vertical tail has the highest contribution, so the C_{y_β} is the determined as:

$$C_{y_\beta} \cong C_{y_{\beta_v}} = -C_{L_{\alpha_v}}\eta_v\left(1 - \frac{d\sigma}{d\beta}\right)\frac{S_v}{S} \tag{4.98}$$

If the disturbance in sideslip angle is positive ($+\Delta\beta$) the reacting side force is desired to be negative (to oppose and nullify the side slip angle). This implies that, the derivative C_{y_β} should be negative to have a positive contribution in dynamic lateral-directional stability. This derivative is an important derivative in shaping characteristics of dutch-roll mode.

4.6.3   Aircraft Reaction to Disturbance in Roll Rate

A roll rate perturbation causes a linear velocity distribution over the wing, vertical tail, and horizontal tail. This consequently induces a local change in angle of attack over each of these surfaces that results in a change in their lift distributions. During any rotation of an object in a circular path, the linear speed (V) at any point away from the center of rotation, is equal to the angular speed (ω) multiplied by the radius of rotation (R); so, V = R ω.

During a roll, for the wing, the linear speed at each wing section is equal to the roll rate (see Fig. 4.11) multiplied by the distance to the x-axis (i.e., p.y_w) along span. When roll rate perturbation is low, the wing local induced angle of attack is:

$$\Delta\alpha_w = \tan^{-1}\left(\frac{p \cdot y_w}{U_1}\right) \approx \frac{p \cdot y_w}{U_1} \qquad (4.99)$$

where y_w is the local distance between each wing section to the aircraft x-axis (i.e., along the span). Similarly, for the vertical tail, the linear speed at each tail section is the roll rate multiplied by the distance to the x-axis (i.e., pz_v) along tail span (z-axis). Thus, when roll rate perturbation is low, the tail induced local angle of attack is:

$$\Delta\alpha_v = \tan^{-1}\left(\frac{pz_v}{U_1}\right) \approx \frac{pz_v}{U_1} \qquad (4.100)$$

The local induced angle of attack on a lifting surface (e.g., wing, tail) generates a local lift, which in turn, can create rolling and yawing moments too.

The disturbance in roll rate (Δp) will cause three separate and parallel aircraft reactions: (1) Aerodynamic rolling moment, (2) Aerodynamic yawing moment, and (3) Aerodynamic side force. These perturbed-state moments and force are discussed here.

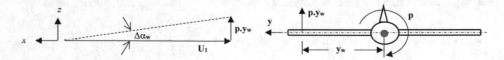

Fig. 4.11 Wing induced angle of attack due to roll rate

a. **Aerodynamic rolling moment due to a roll rate disturbance**

When an aircraft is flying with a steady state angle of attack and zero bank angle (e.g., cruise and climb), a disturbance in roll rate (p) will cause a linear velocity distribution over the wing, vertical tail, and horizontal tail. This consequently causes a local change in angle of attack over each of these surfaces that results in a change in their lift distributions. Thus, these surfaces will generate three perturbed-state rolling moments (L); all in the same direction. This reaction is modeled as the dimensional derivative L_p or the non-dimensional derivative C_{l_p}. The most important non-dimensional lateral dynamic stability derivative is the rate of change of rolling moment coefficient with respect to a disturbance in roll rate; $\frac{\partial C_l}{\partial p}$.

If a disturbance causes the aircraft to experience an increase in roll rate from trim value (+ Δp), a laterally dynamically stable aircraft must develop a negative rolling moment ($-\Delta L$) or a negative rolling moment coefficient ($-\Delta C_l$) to decrease the roll rate to return back to the initial trim value ($p_o = \phi_o = 0$). If the disturbance in roll rate is positive (increases the bank angle), the reacting rolling moment is desired to be negative (to restore the roll rate). This implies that the derivative C_{l_p} should be negative to have a positive contribution in dynamic lateral stability.

$$C_{l_p} = \frac{\partial C_l}{\partial p} \frac{2U_o}{b} < 0 \tag{4.101}$$

The dimensional derivative L_p is obtained by:

$$L_p = \frac{\bar{q}_1 S b^2 C_{l_p}}{2 I_{xx} U_1} \tag{4.102}$$

where non-dimensional derivative C_{l_p} is called *roll damping derivative*, and usually lies in the range -0.8 to -0.1 per radian.

A disturbance in roll rate induces angle of attack along the wing span which in turn, creates an asymmetrical lift distribution which opposes the roll rate. For a fixed-wing aircraft, the *roll damping derivative* is determined by adding wing, fuselage, horizontal tail, and vertical tail contributions:

$$C_{l_p} = C_{l_{p_{wf}}} + C_{l_{p_h}} + C_{l_{p_v}} \tag{4.103}$$

Reference [7] provides the following estimation for this derivative:

$$C_{l_p} = -\frac{C_{L_\alpha}}{12} \frac{1 + 3\lambda}{1 + \lambda} \tag{4.104}$$

where λ is the wing taper ratio (Eq. 2.48). Method for estimating the aircraft components contributions to C_{l_p} can be found in [4] or part VI of Ref. [6].

b. **Aerodynamic yawing moment due to a roll rate angle disturbance**

When an aircraft is flying with a steady state angle of attack and zero bank angle (e.g., cruise), a disturbance in roll rate (p) will cause a perturbed-state yawing moment (N). This reaction is modeled as the dimensional derivative N_p or the non-dimensional derivative C_{n_p}. The dimensional derivative N_p is obtained by:

$$N_p = \frac{\bar{q}_1 S b^2 C_{n_p}}{2 I_{zz} U_1} \tag{4.105}$$

where non-dimensional derivative C_{n_p} is the rate of change of yawing moment coefficient with respect to disturbance in roll rate ($\frac{\partial C_n}{\partial p}$). A disturbance in roll rate induces angle of attack along the wing span, horizontal tail span, and vertical tail span. Since right and left wing-, horizontal tail-, and vertical tail-sections experience opposite effect, a yawing moment is generated due to this perturbation. For a fixed-wing aircraft, the stability *derivative* C_{n_p} is determined by adding wing, fuselage, horizontal tail, and vertical tail contributions:

$$C_{n_p} = \frac{\partial C_n}{\partial p} \frac{2U_o}{b} = C_{n_{p_{wf}}} + C_{n_{p_h}} + C_{n_{p_v}} \tag{4.106}$$

The vertical tail has the biggest contribution, so:

$$C_{n_p} \cong C_{n_{p_v}} = 2 C_{L_{\alpha_v}} \eta_v \left(\frac{l_v}{b} \right) \left(\frac{z_{v_w}}{b} \right) \frac{S_v}{S} \tag{4.107}$$

where z_{v_w} is distance between the vertical tail ac to aircraft cg along the z -axis in wind coordinate system. It may be obtained by:

$$z_{v_w} = z_v \cos(\alpha) \tag{4.108}$$

where z_v is distance between the vertical tail ac to aircraft cg along z body-axis. Note that, if the vertical tail ac is above cg (which is the case for low angle of attacks, α), the z_v is negative.

The effect of this derivative on aircraft lateral-directional dynamic stability is often rather weak (often a very low negative value). Reference [7] provides the following estimation for this derivative:

$$C_{n_p} = -\frac{C_L}{8} \tag{4.109}$$

Method for estimating the aircraft components contributions to C_{n_p} can be found in [4] or part VI of Ref. [6].

c. Aerodynamic side force due to a roll rate disturbance

A disturbance in roll rate (p) to an aircraft—when it is cruising with a level flight—will cause a perturbed-state side force (Y). This reaction is modeled as the dimensional derivative Y_p which is obtained by:

$$Y_p = \frac{\overline{q}_1 S b C_{y_p}}{2 m U_1} \tag{4.110}$$

The non-dimensional derivative C_{y_p} is the side force coefficient due to roll rate derivative $\left(\frac{\partial C_y}{\partial p}\right)$ and is determined by adding the contribution of wing, fuselage, horizontal tail, and vertical tail. However, the wing, fuselage, and horizontal tail contributions tends to be negligible. The vertical tail has the highest contribution, so the C_{y_p} is the determined as:

$$C_{y_p} = \frac{\partial C_y}{\partial p} \frac{2 U_o}{b} \cong C_{y_{p_v}} = -2 C_{L_{\alpha_v}} \eta_v \left(\frac{z_v}{b}\right) \frac{S_v}{S} \tag{4.111}$$

This derivative is often negative at low angles of attack but becomes positive at high angles of attack. The sign of the moment arm, z_v reverses at high angles of attack (i.e., at high angle of attack, the ac_v will be lower than the aircraft cg). In general, this derivative is not an important derivative in terms of its contribution to lateral-directional stability.

4.6.4   Aircraft Reaction to Disturbance in Yaw Rate

The disturbance in yaw rate (Δr) will cause three separate and parallel aircraft reactions: (1) Aerodynamic yawing moment, (2) Aerodynamic rolling moment, and (3) Aerodynamic side force. These perturbed-state moments and force are discussed here. A yaw rate perturbation induces local changes in angle of attack on the vertical tail along its span.

During a yaw, the linear speed at each vertical tail section (see Fig. 4.12) is equal to the yaw rate multiplied by the distance to the z-axis (i.e., r x_v) along vertical tail span. Thus, when yaw rate perturbation is low, the vertical tail induced local angle of attack is:

$$\Delta \alpha_v = \tan^{-1}\left(\frac{r x_v}{U_1}\right) \approx \frac{r x_v}{U_1} \tag{4.112}$$

where x_v is the local distance between each vertical tail section to the aircraft z-axis (i.e., along its span). The local induced angle of attack on the vertical tail generates a local lift, which in turn, can create rolling and yawing moments too.

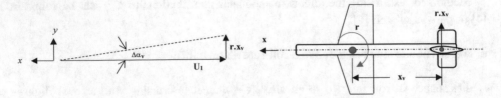

Fig. 4.12 Vertical tail induced angle of attack due to yaw rate

a. Aerodynamic yawing moment due to a yaw rate angle disturbance

When an aircraft is flying with a steady state angle of attack and zero side slip angle (e.g., cruise), a disturbance in yaw rate (r) will cause a perturbed-state yawing moment (N). This reaction is modeled as the dimensional derivative N_r or the non-dimensional derivative C_{n_r}. The most important non-dimensional directional dynamic stability derivative is the rate of change of yawing moment coefficient with respect to disturbance in yaw rate $(\frac{\partial C_n}{\partial r})$.

If a disturbance causes the aircraft to experience an increase in yaw rate from a trim value $(+\Delta r)$, a directionally dynamically stable aircraft must develop a negative yawing moment $(-\Delta N)$ or a negative yawing moment coefficient $(-\Delta C_n)$ to decrease the yaw rate to return back to the initial trim value $(r_o = \beta_o = 0)$. If the disturbance in yaw rate is positive (i.e., increases the side slip angle), the reacting yawing moment is desired to be negative (to restore the side slip angle). This implies that the derivative C_{n_r} should be negative to have a positive contribution in dynamic directional stability.

$$C_{n_r} = \frac{\partial C_n}{\partial r} \frac{2U_o}{b} < 0 \tag{4.113}$$

The dimensional derivative N_r is obtained by:

$$N_r = \frac{\overline{q}_1 S b^2 C_{n_r}}{2 I_{zz} U_1} \tag{4.114}$$

where non-dimensional derivative C_{n_r} is called *yaw damping derivative*. A disturbance in yaw rate induces angle of attack along the vertical tail that creates a vertical tail lift which consequently generates a yawing moment to oppose the yaw rate perturbation. For a fixed-wing the aircraft, the *yaw damping derivative* is determined by adding wing, fuselage, horizontal tail and vertical tail contributions. The wing, horizontal tail, and fuselage contributions at low angles of attack often tend to be negligible. The *yaw damping derivative* is determined as:

$$C_{n_r} \cong C_{n_{r_v}} = -2C_{L_{\alpha_v}} \eta_v \left(\frac{l_v}{b}\right)^2 \frac{S_v}{S} \tag{4.115}$$

Hence, the restoring yawing moment is mainly created by the vertical tail.

b. Aerodynamic rolling moment due to a yaw rate disturbance

When an aircraft is flying with a steady state angle of attack and zero side slip angle (e.g., cruise), a disturbance in yaw rate (r) will cause a perturbed-state rolling moment (L). This reaction is modeled as the dimensional derivative L_r or the non-dimensional derivative C_{l_r}. The dimensional stability derivative L_r is obtained by:

$$L_r = \frac{\bar{q}_1 S b^2 C_{l_r}}{2 I_{xx} U_1} \tag{4.116}$$

where non-dimensional derivative C_{l_r} is the rate of change of yawing moment coefficient with respect to disturbance in yaw rate ($\frac{\partial C_l}{\partial r}$). For a fixed-wing aircraft, the stability derivative C_{l_r} is determined by adding wing, fuselage, horizontal tail, and vertical tail contributions:

$$C_{l_r} = \frac{\partial C_l}{\partial r} \frac{2 U_o}{b} = C_{l_{r_{wf}}} + C_{l_{r_h}} + C_{l_{r_v}} \tag{4.117}$$

The vertical tail has the biggest contribution to this reaction:

$$C_{l_r} \cong C_{l_{r_v}} = 2 C_{L_{\alpha_v}} \eta_v \left(\frac{l_v}{b}\right)\left(\frac{z_v}{b}\right)\frac{S_v}{S} \tag{4.118}$$

where z_v is the distance between vertical tail ac to aircraft cg along z axis.

The restoring rolling moment is mainly created by the vertical tail. The effect of this derivative on aircraft lateral-directional dynamic stability is often weak (a very low positive value).

c. Aerodynamic side force due to a yaw rate disturbance

A disturbance in yaw rate (r) to an aircraft—when it is cruising with a level flight—will cause a perturbed-state side force (Y). This reaction is modeled as the dimensional derivative Y_r which is obtained by:

$$Y_r = \frac{\bar{q}_1 S b C_{y_r}}{2 m U_1} \tag{4.119}$$

The non-dimensional derivative C_{y_r} is the side force coefficient due to yaw rate derivative ($\frac{\partial C_y}{\partial r}$) and is determined by adding the contribution of wing, fuselage, horizontal tail, and vertical tail. However, the wing, fuselage, and horizontal tail contributions tends to be negligible. The vertical tail has the highest contribution, so the C_{y_r} is the determined as:

$$C_{y_r} = \frac{\partial C_y}{\partial r}\frac{2U_o}{b} \cong C_{y_{r_v}} = 2C_{L_{\alpha_v}}\eta_v\left(\frac{l_v}{b}\right)\frac{S_v}{S} \tag{4.120}$$

This derivative is often positive, but has a low value. In general, this derivative is not very important derivative in terms of its contribution to lateral-directional stability.

Example 4.6 Consider a twinjet airliner with a weight of 120,000 lb, a wing span of 95 ft, and a wing area of 1000 ft². The aircraft is cruising with an airspeed of 800 ft/s at the altitude of 30,000 ft where the air density is 8.91×10^{-4} slug/ft³. Because of a sudden gust to the side of the fuselage, the aircraft is experiencing a yaw rate of -10 deg/s for a very short time. The aircraft has the following characteristics:

$$C_{L_{\alpha_v}} = 4.2\frac{1}{\text{rad}}; l_v = 35 \text{ ft}; z_v = 6 \text{ ft}; \eta_v = 0.97; S_v = 220\,\text{ft}^2$$

What rolling and yawing moments are created as the response of the aircraft to this perturbation?

Solution

$$C_{l_r} = 2C_{L_{\alpha_v}}\eta_v\left(\frac{l_v}{b}\right)\left(\frac{z_v}{b}\right)\frac{S_v}{S}$$

$$= 2 \times 4.2 \times 0.97\left(\frac{35}{95}\right)\left(\frac{6}{95}\right)\left(\frac{220}{1000}\right) = 0.042\frac{1}{\text{rad}} \tag{4.118}$$

$$C_{n_r} = -2C_{L_{\alpha_v}}\eta_v\left(\frac{l_v}{b}\right)^2\frac{S_v}{S} = -2 \times 4.2 \times 0.97\left(\frac{35}{95}\right)^2\left(\frac{220}{1000}\right) = -0.243\frac{1}{\text{rad}} \tag{4.115}$$

$$C_{l_r} = \frac{\partial C_l}{\partial r}\frac{2U_o}{b} \Rightarrow C_l = C_{l_r}r\frac{b}{2U_o} = 0.042 \times \frac{-10}{57.3}\frac{95}{2 \times 800} = -0.00043 \tag{4.117}$$

$$C_{n_r} = \frac{\partial C_n}{\partial r}\frac{2U_o}{b} \Rightarrow C_n = C_{n_r}r\frac{b}{2U_o} = -0.243 \times \frac{-10}{57.3}\frac{95}{2 \times 800} = 0.003 \tag{4.113}$$

$$N = \frac{1}{2}\rho V^2 SbC_n = \frac{1}{2} \times 0.00089 \times (800)^2 \times 1000 \times 95 \times 0.003 = 68,220\,\text{Nm} \tag{4.76}$$

$$L = \frac{1}{2}\rho V^2 SbC_l = \frac{1}{2} \times 0.00089 \times (800)^2 \times 1000 \times 95 \times (-0.00043) = -11,690\,\text{Nm} \tag{4.80}$$

4.6.5 Aircraft Reaction to a Sudden Change in Heading Angle

In a symmetric aircraft, a disturbance in the heading angle—if sideslip angle is not disturbed—will not generate any aircraft reaction in terms of side force, rolling moment and yawing moment. Thus, an aircraft does not feature a heading stability, and does not have any memory of its heading angle. If a gust changes only the heading angle, (e.g., in a pure yawing motion), the aircraft does not have a tendency to return to its initial heading angle (i.e., an aircraft has neutral stability in heading). The restoration of the heading angle due to a disturbance requires an automatic closed-loop control system (either by human pilot or autopilot).

4.7 Contributions of Aircraft Components to Lateral-Directional Stability

Every aircraft component (e.g., wing, fuselage, tail, and propulsion unit) has a unique contribution to the aircraft overall lateral-directional stability. An aircraft is laterally-directionally either stable, or unstable, or neutrally stable; however, a component may have a positive contribution (be stabilizing) or a negative contribution (be destabilizing). In this section, contributions of aircraft components to lateral-directional stability will be discussed.

These discussions aim to demonstrate the relationship between the lateral-directional stability derivatives and the geometric and configuration characteristics of the aircraft. An important application of this information is during aircraft design process, which helps the designer to select the best configuration to meet lateral-directional handling quality requirements.

Main component contribution is explored with respect to the role each play in generating a restoring rolling moment (for lateral stability), and a restoring yawing moment (for directional stability). In general, wing has the greatest contribution to lateral stability, while vertical tail has the highest contribution to directional stability.

To analyze aircraft components contributions to lateral static stability, the focus is on the role of each component (1. positive/negative, and 2. amount) in lateral static stability derivative, C_{l_β}. To analyze aircraft components contributions to lateral dynamic stability, the focus is on the role of each component (1. positive/negative, and 2. amount) in roll damping derivative, C_{l_p}.

To analyze aircraft components contributions to directional static stability, the focus is on the role of each component (1. positive/negative, and 2. amount) in directional static stability derivative, C_{n_β}. To analyze aircraft components contributions to directional dynamic stability, the focus is on the role of each component (1. positive/negative, and 2. amount) in yaw damping derivative, C_{n_r}.

4.7.1 Wing Contribution

The wing has the greatest contribution to lateral stability, while has a minor effect on directional stability. Three parameters of a wing play significant role in the lateral stability: (1) Dihedral angle (Γ), (2) Sweep angle (Λ), and (3) Wing vertical position on the fuselage. The mechanism and impact of these three parameters on the wing contribution to lateral stability are discussed in the following.

4.7.1.1 Dihedral Angle

In a fixed-wing aircraft, the lateral stability is mainly provided by the wing dihedral angle (Γ). The dihedral angle is defined as the spanwise inclination of the wing with respect to the horizontal (viewed from front-view). If the tip section is higher than the root section, the angle is positive ($\Gamma > 0$); if the tip section is lower than the root section, the angle is negative ($\Gamma < 0$, which is commonly called anhedral).

Consider the aircraft in Fig. 4.13 that due to a gust, it is disturbed from a wing-level attitude. The gust makes the aircraft to have a positive roll (right-wing down and left-wing up) and has a bank angle, ϕ. Then, aircraft begins to sideslip to the right with a side velocity of v, so the relative wind is coming from right-side. The sideslip motion will create a sideslip angle of β where:

$$\beta = \sin^{-1}(v/U_o) \tag{4.121}$$

Due to dihedral angel, normal component of side velocity is

$$v_n = v\sin(\Gamma) \approx v\Gamma \tag{4.122}$$

The right-wing section (leading) experiences an increased angle of attack and consequently an increase in lift. The left-wing section (trailing) experiences the opposite effect.

$$\Delta\alpha = \sin^{-1}\left(\frac{v_n}{U_o}\right) \tag{4.123}$$

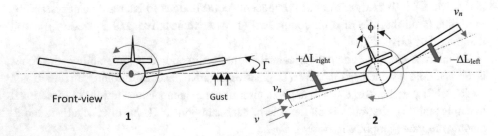

Fig. 4.13 Impact of the wing dihedral angle to lateral stability

By assuming small sideslip angle and small angle of attack, we approximate (i.e., linearize):

$$\beta \approx v/U_o \tag{4.124}$$

$$\Delta\alpha \approx \frac{v_n}{U_o} \tag{4.125}$$

By inserting Eqs. (4.122) and (4.124) into Eq. (4.125), we obtain:

$$\Delta\alpha = \frac{v\Gamma}{U_o} = \beta\Gamma \tag{4.126}$$

Thus,

$$\Delta\alpha_{right} \approx \beta\Gamma \tag{4.127}$$

$$\Delta\alpha_{left} \approx -\beta\Gamma \tag{4.128}$$

The sideslipping motion will create an additional corresponding lift $(+\Delta L)$ on the downward-moving wing, while will create a negative corresponding lift $(-\Delta L)$ on the upward-moving wing. This lift couple will generate a negative (i.e., restoring) rolling moment, so the contribution of dihedral is stabilizing. The greater the dihedral angel, the C_{l_β} would be more negative.

As the dihedral angle is increased, the restoring rolling moment will be greater. Hence, the greater the dihedral angle, the more stabilizing will be a wing. However, as the dihedral angle is increased, the total lift created by the wing will be decreased (an undesired consequence). For a given angle of attack, a decrease in dihedral angle will result in a higher lift coefficient. Therefore, the dihedral angle should be optimized to satisfy both lift requirements as well as the lateral stability requirements.

4.7.1.2 Wing Vertical Position on the Fuselage

The contribution of wing vertical position on the fuselage to lateral stability is illustrated in Fig. 4.14. The aircraft due to a gust is disturbed from a wing-level attitude, so it will have a positive roll (right-wing down and left-wing up) and. Then, aircraft begins to sideslip to the right, so the relative wind is coming from right-side. The cross-flow in sideslip is split into two parts: (1) a part on top of the wing, (2) a part under the wing.

That part of the side-flow which hits the fuselage, has to turn around it, and move along the shortest path. In a high-wing configuration, the shortest distance for the flow under the wing is to turn up above the wing; then, move over the wing, and then, turn down again to continue the path. In a low-wing configuration, the shortest distance for the flow on top of the wing is to turn down under the wing, then, move along the wing, and then, turn up again to continue the motion.

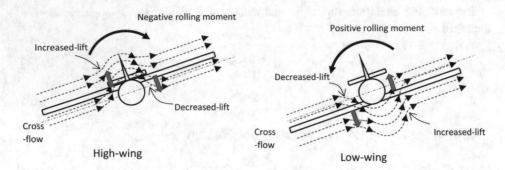

Fig. 4.14 Wing vertical position on the fuselage contribution (front-view)

A turn of the flow at the inboard wing stations creates a local change in wing angle of attack. When the flow turns down, it will induce a negative angle of attack (i.e., a negative local lift; $-\Delta L$); while, when the flow turns up, it will induce a positive angle of attack (i.e., an additional local lift; $+\Delta L$).

In a high-wing, the lift couple will generate a negative (i.e., restoring) rolling moment so the contribution is stabilizing (i.e., $C_{l_\beta} < 0$). In a low-wing, the lift couple will generate a positive rolling moment so the contribution is destabilizing. A high-wing makes the C_{l_β} more negative, while a low-wing makes the C_{l_β} less negative.

A combination of wing dihedral and a high-wing will have a stronger stabilizing effect. To balance the stabilizing level, a high-wing has considerably less geometric dihedral than in a low wing aircraft. To maintain the same C_{l_β} configuration, a low-wing requires a significantly greater dihedral angle than a high-wing configuration. Sometimes, an anhedral (i.e., negative dihedral) is recommended for a high-wing, to satisfy lateral-directional stability requirements.

For instance, the Boeing 747-8 (Fig. 4.15a) transport aircraft with a low wing configuration has a $+7°$ of dihedral angle, while the MD/BAe Harrier II (Fig. 4.15b) V/STOL Close Support with a high wing configuration has a $-14.5°$ of dihedral angle.

Fig. 4.15 Boeing 747-8 and MD/BAe Harrier II

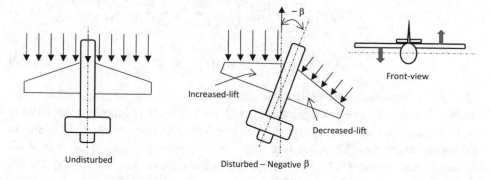

Fig. 4.16 Wing sweep angle contribution (top-view, front-view)

4.7.1.3   Sweep Angle

Basically, wing sweep angle improves static lateral stability, and impacts directional stability. The contribution of wing sweep angle to lateral stability is illustrated in Fig. 4.16, where a trimmed aircraft sideslip angle (i.e., $\beta = 0$) is disturbed. After perturbation (here, $\beta < 0$), the windward (i.e., left) wing section has an effective decrease in sweep angle, while the trailing (right) wing section experiences an effective increase in sweep angle. This implies that, there is a difference in velocity components normal to the leading edge between the left- and right-wing sections. Thus, the windward wing (with a less effective sweep) will experience more lift than the other side. This lift couple will generate a positive (i.e., restoring) rolling moment so the contribution is stabilizing (i.e., $C_{l_\beta} < 0$).

The sweep angle reinforces the dihedral effect and makes the dihedral effect (C_{l_β}) more negative. Hence, a swept wing may not need a dihedral to satisfy lateral stability requirements. When wing sweep is large, the aircraft is laterally too stable (i.e., C_{l_β} is too negative). One solution is to have negative dihedral on wing or horizontal tail to offset this effect (e.g., McDonnell Douglas F-4 Phantom II and McDonnell Douglas AV-8B Harrier II).

However, making the dihedral effect (C_{l_β}) more negative will make an aircraft more spirally stable. At the same time, the dutch roll damping ratio tends to decrease. This presents a design conflict which must be resolved through some compromise.

4.7.2   Vertical Tail Contribution

The vertical tail has the highest contribution to directional stability (both static and dynamic). To analyze vertical tail contribution to directional static stability, we focus on the role it plays (1. positive/negative, and 2. amount) in directional static stability derivative, C_{n_β} and yaw damping derivative; C_{n_r}. For convenience, both equations are repeated:

$$C_{n_\beta} \cong C_{n_{\beta_v}} = C_{L_{\alpha_v}}\left(1 - \frac{d\sigma}{d\beta}\right)\eta_v\frac{S_v l_v}{Sb} \tag{4.78}$$

$$C_{n_r} \cong C_{n_{r_v}} = -2C_{L_{\alpha_v}}\eta_v\left(\frac{l_v}{b}\right)^2\frac{S_v}{S} \tag{4.115}$$

To have a directionally statically stable aircraft, the C_{n_β} should be positive. Moreover, to have a directionally dynamically stable aircraft, the C_{n_r} should be negative. All terms in Eqs. (4.78) and (4.115) have a negative sign, which implies that a vertical tail is always directionally stabilizing. There are four terms that represent the amount of vertical tail stabilizing contribution: (1) Tail area (S_v), (2) Tail arm ($l_v = x_{ac_v} - x_{cg}$), (3) Tail lift curve slope ($C_{L_{\alpha_v}}$), and (4) Tail efficiency (η_v).

The first two parameters can be integrated in the vertical tail volume ratio (V_V). The tail lift curve slope can be increased by proper design/selection of its airfoil, plus increasing tail aspect ratio. As the vertical tail area, vertical tail arm, vertical tail lift curve slope, and vertical tail efficiency are increased, the vertical tail positive contribution is greater. Moreover, the side force on the vertical tail produces both a yawing moment and a rolling moment.

For an aircraft with a V-tail or Y-tail (e.g., General Atomics MQ-1 Predator UAV), tail area (S_v) is the vertical projection of the total planform area of the V-tail (S_{vt}):

$$S_v = S_{vt}sin(\Gamma_{vt}) \tag{4.129}$$

where Γ_{vt} is the V-tail dihedral angle (viewed from front view). In the Predator UAV, tail dihedral angle of the top portion is estimated to be 32°.

The vertical tail contribution to the static directional stability of the aircraft can be tailored by proper selection of V_V and $C_{L_{\alpha_v}}$. We can make a similar conclusion for the contribution of vertical tail to dynamic directional stability by considering Eq. (4.115). One low-cost method to improve directional stability is to increase the vertical tail area by adding a *dorsal fin* (on top) or *ventral fin* (under the rear fuselage).

The vertical tail contribution to lateral stability is produced by the tail lift due to sideslip. The rolling moment occurs because the vertical tail aerodynamic center is often located above the aircraft center of gravity. The rolling moment produced by the vertical tail tends to bring the aircraft back to a wing-level attitude. Thus, the vertical tail is laterally stabilizing.

4.7.3 Horizontal Tail Contribution

The contribution of a horizontal tail to the lateral stability and dihedral effect (C_{l_β}) is very similar to the contribution of the wing, but just with a smaller scale. The horizontal tail contribution in a symmetric configuration aircraft has minor contribution to the directional stability and is negligible. However, since horizontal tail is in the same location of the vertical tail (in a conventional tail configuration), there is an indirect minor impact.

4.7.4 Fuselage Contribution

The fuselage contribution to lateral stability (via C_{l_β}) is illustrated in Fig. (4.14). The fuselage in a high-wing configuration makes the C_{l_β} more negative, while at a low-wing makes the C_{l_β} less negative. Thus, in a high-wing configuration, the fuselage is stabilizing, while at a low-wing is destabilizing.

The fuselage contribution to directional stability (via C_{n_β}) is represented in Eq. (4.74) as $\frac{d\sigma}{d\beta}$. The fuselage geometry and configuration will change the sidewash angle and sidewash ratio at the vertical tail. As the value of $\frac{d\sigma}{d\beta}$ is decreased, the fuselage is directionally stabilizing, while it is increased, the fuselage is directionally destabilizing. Technique to estimate $\frac{d\sigma}{d\beta}$ is presented in [4].

4.7.5 Aircraft Weight, Moment of Inertia, and Center of Gravity Contribution

Aircraft weight, aircraft mass moment of inertia about x- and z-axes (i.e., I_{xx} and I_{zz}), and aircraft center of gravity have direct contributions to lateral and directional stability respectively. Since every component has weight, center of gravity, and geometry; all aircraft components have a direct contribution to static and dynamic lateral-directional stability. Any change in weight, center of gravity, and geometry of a component will directly impact longitudinal stability. Hence, one way to improve lateral-directional stability is to change the components' locations and the weight distribution.

As the aircraft center of gravity is moved forward, the aircraft is directionally more stable. The more forward the cg, the component contribution become more stabilizing.

The mass moment of inertias about x- and z-axes (i.e., I_{xx} and I_{zz}) are a function of components weight, geometries, and their distances to x- and z-axes. As the aircraft weight and I_{xx} and I_{zz} are increased (keeping the same geometry), the aircraft become laterally/directionally more stable.

4.8   Problems

1. A GA aircraft has the following lateral-directional characteristic equation for a cruising flight:

$$207s^4 + 3045s^3 + 7426s^2 + 38{,}587s - 337.9 = 0$$

Identify the dutch roll, roll, and spiral modes of lateral-directional dynamics; and then determine damping ratio, natural frequency, and period of the dutch roll mode, and time constants of roll mode and spiral mode. Is the aircraft dynamically lateral-directionally stable? Why?

2. A fighter aircraft has the following lateral-directional characteristic equation for a cruising flight:

$$887s^4 + 1445s^3 + 5326s^2 + 6887s + 89 = 0$$

Identify the dutch roll, roll, and spiral modes of lateral-directional dynamics; and then determine damping ratio, natural frequency, and period of the dutch roll mode, and time constants of roll mode and spiral mode. Is the aircraft dynamically lateral-directionally stable? Why?

3. A jet fighter aircraft with a mass of 8000 kg, wing area of 20 m^2 and wing span of 7 m is cruising at 40,000 ft with a speed of Mach 1.8. The lateral-directional characteristic equation of the aircraft is as follows:

$$860s^4 + 830s^3 + 1574s^2 + 1743s + 50 = 0$$

a. Is this aircraft lat-dir dynamically stable?
b. Identify the modes and their characteristics, damping ratio, and un-damped natural frequency (for second order mode), and time constant (for first order modes).

4. A twin-engine GA aircraft has the following sideslip-angle-to-rudder-deflection transfer function for a cruising flight:

$$\frac{\beta(s)}{\delta_R(s)} = \frac{801s^2 + 362s + 243}{167s^4 + 384s^3 + 860s^2 + 12430s - 22}$$

5. Is the aircraft dynamically laterally-directionally stable? Why?

A large transport aircraft has the following sideslip-angle-to-aileron-deflection transfer function for a cruising flight:

$$\frac{\beta(s)}{\delta_A(s)} = \frac{-6s^2 + 4s + 0.9}{832s^4 + 630s^3 + 780s^2 + 440s + 5.2}$$

Is the aircraft dynamically laterally-directionally stable? Why?

6. Consider a fixed-wing General Aviation aircraft with a maximum takeoff weight of 2720 lb, a wing planform area of 183 ft^2, and the following characteristics:

$$\frac{d\sigma}{d\beta} = 0.24; \ C_{L_{\alpha_v}} = 4.5 \ \frac{1}{rad}; \ b = 36 \ \text{ft}; \ \eta_v = 0.97; \ S_v = 38 \ \text{ft}^2; \ l_v = 20.1 \ \text{ft}$$

Determine directional static statically derivative; C_{n_β}. Is this aircraft directionally statically stable?

7. Consider a large transport aircraft with a maximum takeoff weight of 700,000 lb, a wing planform area of 4200 ft², and the following characteristics:

$$\frac{d\sigma}{d\beta} = 0.23; \; C_{L_{\alpha_v}} = 4.6\,\frac{1}{\text{rad}}; \; b = 202\,\text{ft}; \; \eta_v = 0.95; \; S_v = 832\,\text{ft}^2; \; l_v = 104\,\text{ft}$$

Determine directional static statically derivative; C_{n_β}. Is this aircraft directionally statically stable?

8. A large transport aircraft has the following lateral-directional state space model for a cruising flight at M = 0.87 at 32,000 ft:

$$\begin{bmatrix} \dot{\beta} \\ \dot{p} \\ \dot{r} \\ \dot{\phi} \end{bmatrix} = \begin{bmatrix} 0 & 0.03 & -1.3 & -0.4 \\ -2.4 & 0 & 0.04 & -15 \\ 0.035 & 0 & -0.3 & 8.3 \\ 0 & 1 & 0 & 0 \end{bmatrix} \begin{bmatrix} \beta \\ p \\ r \\ \phi \end{bmatrix} + \begin{bmatrix} 0 & 0.01 \\ 4 & 4.7 \\ -0.1 & -2.1 \\ 0 & 0 \end{bmatrix} \begin{bmatrix} \delta_A \\ \delta_R \end{bmatrix}$$

$$\begin{bmatrix} \beta \\ p \\ r \\ \phi \end{bmatrix} = \begin{bmatrix} 1 & 0 & 0 & 0 \\ 0 & 1 & 0 & 0 \\ 0 & 0 & 1 & 0 \\ 0 & 0 & 0 & 1 \end{bmatrix} \begin{bmatrix} \beta \\ p \\ r \\ \phi \end{bmatrix} + \begin{bmatrix} 0 & 0 \\ 0 & 0 \\ 0 & 0 \\ 0 & 0 \end{bmatrix} \begin{bmatrix} \delta_A \\ \delta_R \end{bmatrix}$$

Is the aircraft dynamically laterally-directionally stable? Why?

9. The matrix A of lateral-directional state space model for a business jet aircraft at a cruising flight at M = 0.5 at 25,000 ft:

$$A_{lat-dir} = \begin{bmatrix} -3.5 & -4.1 & -12 & -0.3 \\ 1 & 0 & 0 & 0 \\ 0 & 1 & 0 & 0 \\ 0 & 0 & 1 & 0 \end{bmatrix}$$

Is the aircraft dynamically laterally-directionally stable? Why?

10. The damping ratio and natural frequency of the dutch roll mode of a business jet aircraft at sea level is 0.34 and 0.42 rad/s respectively. Determine the time for halving the amplitude.

11. Consider a twinjet airliner with a weight of 110,000 lb, a wing span of 87 ft, and a wing area of 950 ft². The aircraft is cruising with an airspeed of 750 ft/s at the altitude of 30,000 ft. Because of a sudden gust to the side of the fuselage, the aircraft is experiencing a yaw rate of − 6 deg/s for a very short time. The aircraft has the following two stability derivatives: $C_{l_r} = 0.24\,\frac{1}{\text{rad}}$; $C_{n_r} = -0.22\,\frac{1}{\text{rad}}$.

What rolling and yawing moments are created as the response of the aircraft to this perturbation?

12. Consider a twin-prop aircraft with a weight of 15,000 lb, a wing span of 87 ft, and a wing area of 340 ft^2. The aircraft is cruising with an airspeed of 55 ft/s at the altitude of 20,000 ft where the air density is 12.67×10^{-4} slug/ft^3. Because of a sudden gust to the side of the fuselage, the aircraft is experiencing a yaw rate of $-$ 12 deg/s for a very short time. The aircraft has the following two stability derivatives: $C_{l_r} = 0.23 \frac{1}{\text{rad}}; C_{n_r} = -0.21 \frac{1}{\text{rad}}$.

    What rolling and yawing moments are created as the response of the aircraft to this perturbation?

13. Consider a twinjet airliner with a weight of 140,000 lb, a wing span of 104 ft, and a wing area of 1100 ft^2. The aircraft is cruising with an airspeed of 750 ft/s at the altitude of 30,000 ft. Because of a sudden gust to the side of the fuselage, the aircraft is experiencing a yaw rate of $-$ 10 deg/s for a very short time. The aircraft has the following characteristics:

$$CL_{\alpha_v} = 4.4 \frac{1}{\text{rad}}; l_v = 42 \text{ ft}; z_v = 6.5 \text{ ft}; \eta_v = 0.94; S_v = 260 \text{ ft}^2$$

    What rolling and yawing moments are created as the response of the aircraft to this perturbation?

14. A light transport aircraft with a mass of 5000 kg, wing area of 30 m^2 and wing span of 15 m is cruising at sea level with a speed of 230 knot. Data of the aircraft is as follows:

$$C_{y_\beta} = -0.3 \frac{1}{\text{rad}}; C_{l_p} = -0.42 \frac{1}{\text{rad}}; I_{xx} = 11,000 \text{ kg m}^2; I_{zz} = 4000 \text{ kg m}^2$$

    Calculate the following two dimensional derivatives: Y_β, L_P.

15. A large transport aircraft with a mass of 300,000 kg, wing area of 550 m^2 and wing span of 62 m is cruising at 30,000 ft with a speed of 300 knot. Data of the aircraft is as follows:

$$C_{n_\beta} = 0.3 \frac{1}{\text{rad}}; C_{n_p} = -0.12 \frac{1}{\text{rad}}; I_{zz} = 6.7 \times 10,000 \text{ kg m}^2$$

    Calculate the following dimensional directional derivatives N_β and N_P.

16. A jet airliner with the following characteristics is cruising with a speed of 170 knot at sea level.

$$S = 22 \text{ m}^2; S_v = 4 \text{ m}^2; b = 14 \text{ m}; l_v = 7 \text{ m};$$

$$d\sigma/d\beta = 0.2; CL_{\alpha_v} = 4.3 \frac{1}{\text{rad}}; \eta_v = 0.95$$

    A 20 knot wind gust is hitting from right side, and imposes a sideslip angle. What yawing moment is generated by the aircraft as an inherent response to this disturbance?

17. The following jet aircraft is cruising at sea level with a speed of 300 knot. A gust hit to the vertical tail and create a positive yaw rate of 20 deg/s. Determine the Yawing moment and side force response of the aircraft to this perturbation.

$$S = 100\,\text{m}^2;\ S_v = 28\ \text{m}^2;\ b = 30\ \text{m};$$

$$l_v = 11\ \text{m};\ C_{L_{\alpha_v}} = 4.5\frac{1}{\text{rad}};\ \eta_v = 0.94$$

18. The following aircraft is cruising with a speed of 170 knot at 20,000 ft. A gust hits to the fuselage nose from right side and produces a perturbation in roll rate by the magnitude of 17 deg/s. The aircraft has a mass of 50,000 kg, wing area of 300 m², AR of 12, and the lift curve slopes of the wing and tails are 5.5 1/rad. Assume the dynamic pressure ratio at the vertical tail is 0.96 and the vertical tail area 70 m² and $C_{Lo} = 0$, $\delta_E = 0$. Calculate the side force developed by the aircraft in response to this roll rate.

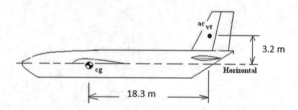

References

1. Kenneth W. Iliff, Richard E. Maine, and T. D. Montgomery, Important Factors in the Maximum Likelihood Analysis of Flight Test Maneuvers, Technical Paper 1459, Scientific & Technical Information Office, National Aeronautics & Space Administration, 1979
2. Steve Klausmeyer, Caleb Fisher, and Kelly Laflin, Stability Derivative Estimation: Methods and Practical Considerations for Conventional Transonic Aircraft, Applied Aerodynamics Conference, Atlanta, Georgia, June 25–29, 2018
3. Roskam J., Airplane flight dynamics and automatic flight controls, 2007, DARCO
4. Hoak D. E., Ellison D. E., et al, USAF Stability and Control DATCOM, Flight Control Division, Air Force Flight Dynamics Laboratory, Wright-Patterson AFB, Ohio, 1978
5. MIL-STD-1797, Flying Qualities of Piloted Aircraft, Department of Defense, Washington DC, 1997
6. Roskam J., Airplane Design, 2015, DAR Corp
7. Nelson R., Flight Stability and Automatic Control, McGraw Hill, 1997

Longitudinal Control

5

5.1 Introduction

Control is defined as to change the aircraft flight conditions from one trim point to another trim point with a desired rate or a specific duration. Longitudinal control is defined as to changing the flight condition from one longitudinal trim condition (e.g., cruise with an initial angle of attack) to another trim condition (e.g., climb with a climb angle) with a desired rate. Longitudinal control is within the xz plane, and the most challenging part is the pitch control (i.e., rotation about y-axis).

An aircraft must be longitudinally controllable at all flight conditions within the flight envelope. Missiles and fighter aircraft are longitudinally highly controllable, while aircraft such as hang gliders are lightly controlled. However, balloons and airships are not aerodynamically controllable, but the altitude can be increased by dropping dead weights or generating more hot air.

In a fixed-wing aircraft, there are mainly two longitudinal control tools: (1) Elevator (δ_E), and (2) Engine throttle (δ_T). The first one is a control surface and is a tool for aerodynamic control, since it creates aerodynamic forces and moment. Moreover, there are two secondary control surfaces for longitudinal aerodynamic control: (1) High lift device (e.g., flap), (2) Spoiler. A spoiler is a mean to produce drag while disrupts (decreases) the lift. Flap increases lift, while adding drag.

In a cruising flight, to control airspeed, there are two main tools: (1) Engine throttle to change airspeed, and (2) Elevator to change pitch angle. However, in practice, when engine throttle is rotated, the engine thrust is changed; thus, the airspeed, pitch angle, and altitude will vary simultaneously.

The forces acting on an aircraft in flight consist of aerodynamic, thrust and gravitational forces. Any change in any of these forces will disturb the trim condition and can be employed as a method to control the aircraft. For instance, while fuel is burning during

© The Author(s), under exclusive license to Springer Nature Switzerland AG 2022   159
M. H. Sadraey, *Flight Stability and Control*, Synthesis Lectures on Mechanical
Engineering, https://doi.org/10.1007/978-3-031-18765-0_5

flight, the control is needed to maintain the flight status (e.g., keep altitude or speed). Controllability and maneuverability specifications are governed by mission requirements, which are defined within handling qualities.

Two longitudinal aerodynamic forces are lift and drag, and one aerodynamic longitudinal moment is the pitching moment. The aerodynamic control of an air vehicle about the pitch axis is accomplished by horizontal tail via elevator deflection. The longitudinal forces and moment created by the elevator deflection is fed back to the pilot via stick.

All longitudinal flight conditions which an aircraft may encounter must be investigated, so that the elevator can be designed to the proper geometry. The longitudinal control gets harder, as: (1) Aircraft gets heavier (keeping geometry), (2) Aircraft cg moves forward, and (3) Aircraft mass moment of inertia about y-axis (I_{yy}) is increased.

In this chapter, fundamentals, parameters, and governing equations for the aircraft longitudinal control are presented. The longitudinal aerodynamic control for a fixed-wing aircraft via elevator, and rotary-wing aircraft via propeller rotation are discussed. Other topics including stick force, stick-fixed, stick-free, and automatic flight control system are explored too.

5.2   Longitudinal Governing Equations of Motion

Five main flight parameters to control in a longitudinal motion are: (1) Airspeed (u), (2) Angle of attack (α), (3) Pitch angle (θ), (4) Pitch rate (q), and (5) Altitude (h). The following differential equations govern the variations of these parameters in a pure longitudinal motion:

$$m\dot{u} = \mathrm{X} + T - mg\cos\Theta_1\theta \tag{5.1}$$

$$m(\dot{w} - U_1 q) = -mg(\sin\Theta_1\theta) + \mathrm{Z} \tag{5.2}$$

$$I_{yy}\dot{q} = M \tag{5.3}$$

$$\dot{h} = u.\cos(\theta) \tag{5.4}$$

$$\dot{\theta} = q \tag{5.5}$$

The first three equations have been derived in Chap. 3. Equation (5.4) demonstrates the definition of rate of climb that govern the variations of altitude. Equation (5.5) presents the definition of pitch rate.

As shown in Chap. 3, the longitudinal aerodynamic forces and pitching moment coefficients are modelled with respect to the dimensional stability and control derivatives. In terms of derivatives, Eqs. (5.1)–(5.3) are expressed as:

$$\dot{u} = -g\theta\cos\Theta_o + X_u u + X_\alpha\alpha + X_{\delta_E}\delta_E + T_{\delta_T}\delta_T \tag{5.6}$$

$$U_o\dot{\alpha} - U_o\dot{\theta} = -g\theta(\sin\Theta_o) + Z_u u + Z_\alpha\alpha + Z_{\dot{\alpha}}\dot{\alpha} + Z_q q + Z_{\delta_E}\delta_E \tag{5.7}$$

$$\dot{q} = M_u u + M_\alpha\alpha + M_{\dot{\alpha}}\dot{\alpha} + M_q q + M_{\delta_E}\delta_E \tag{5.8}$$

where pitch angular acceleration is:

$$\dot{q} = \ddot{\theta} \tag{5.9}$$

where aerodynamic forces along x and z axes are expressed as:

$$X = -D \tag{5.10}$$

and

$$Z = -L \tag{5.11}$$

The aerodynamic forces (lift, L and drag, D) coefficients and pitching moment (M) coefficient are modeled as:

$$C_L = C_{L_o} + C_{L_\alpha}\alpha + C_{L_q}Q\frac{C}{2U_1} + C_{L_{\dot{\alpha}}}\dot{\alpha}\frac{C}{2U_1} + C_{L_u}\frac{u}{U_1} + C_{L_{\delta_e}}\delta_e \tag{5.12}$$

$$C_D = C_{D_o} + C_{D_\alpha}\alpha + C_{D_q}Q\frac{C}{2U_1} + C_{D_{\dot{\alpha}}}\dot{\alpha}\frac{C}{2U_1} + C_{D_u}\frac{u}{U_1} + C_{D_{\delta_e}}\delta_e \tag{5.13}$$

$$C_m = C_{m_o} + C_{m_\alpha}\alpha + C_{m_q}Q\frac{C}{2U_1} + C_{m_{\dot{\alpha}}}\dot{\alpha}\frac{C}{2U_1} + C_{m_u}\frac{u}{U_1} + C_{m_{\delta_e}}\delta_e \tag{5.14}$$

The open-loop control of an aircraft can be modeled by a block (see Fig. 5.1) with multiple inputs and multiple outputs (MIMO). The block contains the aircraft longitudinal dynamic model with either: (1) Transfer function, (2) State space representation. Three aircraft flight angles are controlled in the longitudinal control: (1) Angle of attack, (2) Climb angle, and (3) Pitch angle. Other flight parameters to be controlled in longitudinal motion include: (1) Pitch rate, (2) Airspeed, (3) Altitude, and (4) Rate of climb/descent.

Equations (5.6)–(5.9) can be expressed (derivation is shown in Chap. 3) into state-space model as:

Fig. 5.1 Longitudinal open-loop control system

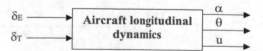

$$
\begin{bmatrix} \dot{u} \\ \dot{\alpha} \\ \dot{q} \\ \dot{\theta} \end{bmatrix} = \begin{bmatrix} X_u & X_\alpha & 0 & -g\cos\Theta_o \\ \frac{Z_u}{U_o} & \frac{Z_\alpha}{U_o} & 1 + \frac{Z_q}{U_o} & \frac{-g(\sin\Theta_o)}{U_o} \\ M_u & M_\alpha & M_q & 0 \\ 0 & 0 & 1 & 0 \end{bmatrix} \begin{bmatrix} u \\ \alpha \\ q \\ \theta \end{bmatrix} + \begin{bmatrix} X_{\delta_E} & T_{\delta_T} \\ Z_{\delta_E} & Z_{\delta_T} \\ M_{\delta_E} & M_{\delta_T} \\ 0 & 0 \end{bmatrix} \begin{bmatrix} \delta_E \\ \delta_T \end{bmatrix} \qquad (5.15)
$$

The output variables are tentatively selected to be: (1) Linear airspeed (u), (2) Angle of attack (α), (3) Pitch rate (q), and (4) climb angle (γ). This, the output equation is presented as:

$$
\begin{bmatrix} u \\ \alpha \\ q \\ \gamma \end{bmatrix} = \begin{bmatrix} 1 & 0 & 0 & 0 \\ 0 & 1 & 0 & 0 \\ 0 & 0 & 1 & 0 \\ 0 & -1 & 0 & 1 \end{bmatrix} \begin{bmatrix} u \\ \alpha \\ q \\ \theta \end{bmatrix} + \begin{bmatrix} 0 & 0 \\ 0 & 0 \\ 0 & 0 \\ 0 & 0 \end{bmatrix} \begin{bmatrix} \delta_E \\ \delta_T \end{bmatrix} \qquad (5.16)
$$

In this longitudinal state space model, A is a 4×4 square matrix, the C is a 4×4 matrix, and B and D are 4×2 matrices, where in D, all elements are often zero.

References [1, 2] has provided longitudinal-lateral-directional stability and control derivatives (subsonic database) of fighter aircraft General Dynamics F-16 Fighting Falcon (Fig. 5.2) based on wind tunnel test results at NASA Langley and Ames Research center. The following is the longitudinal state-space model for only one input (δ_E), and two outputs (q and α):

$$
\begin{bmatrix} \dot{q} \\ \dot{\alpha} \end{bmatrix} = \begin{bmatrix} -0.772 & -1.012 \\ 0.927 & -0.574 \end{bmatrix} \begin{bmatrix} q \\ \alpha \end{bmatrix} + \begin{bmatrix} -3.635 \\ -0.078 \end{bmatrix} \delta_E
$$

This model may be used to simulate and analyze the pitch response of the aircraft to an elevator input.

By applying LaPlace transform to Eqs. (5.6)–(5.9), we can derive the longitudinal transfer functions in s-domain. The airspeed-to-elevator-deflection transfer function is obtained as:

$$
\frac{u(s)}{\delta_E(s)} = \frac{A_u s^3 + B_u s^2 + C_u s + D_u}{A_1 s^4 + B_1 s^3 + C_1 s^2 + D_1 s + E_1} \qquad (5.17)
$$

The angle-of-attack-to-elevator-deflection transfer function is obtained as:

$$
\frac{\alpha(s)}{\delta_E(s)} = \frac{A_\alpha s^3 + B_\alpha s^2 + C_\alpha s + D_\alpha}{A_1 s^4 + B_1 s^3 + C_1 s^2 + D_1 s + E_1} \qquad (5.18)
$$

The pitch-angle-to-elevator-deflection transfer function is obtained as:

$$
\frac{\theta(s)}{\delta_E(s)} = \frac{A_\theta s^2 + B_\theta s + C_\theta}{A_1 s^4 + B_1 s^3 + C_1 s^2 + D_1 s + E_1} \qquad (5.19)
$$

Fig. 5.2 Fighter aircraft General Dynamics F-16 Fighting Falcon

Since pitch rate (q) is the determined by differentiation of pitch angle with time $\left(q(t) = \frac{d\theta(t)}{dt}\right)$, the pitch-rate-to-elevator-deflection transfer function is obtained by multiplying numerator of $\frac{\theta(s)}{\delta_E(s)}$ by an "s" (i.e., in s-domain, $Q(s) = s\theta(s)$). Thus, the pitch-rate-to-elevator-deflection transfer function is:

$$\frac{Q(s)}{\delta_E(s)} = \frac{s\left(A_\theta s^2 + B_\theta s + C_\theta\right)}{A_1 s^4 + B_1 s^3 + C_1 s^2 + D_1 s + E_1} = \frac{A_\theta s^3 + B_\theta s^2 + C_\theta s}{A_1 s^4 + B_1 s^3 + C_1 s^2 + D_1 s + E_1}$$

(5.20)

All longitudinal transfer functions have the same fourth-order denominator (i.e., characteristic equation). The coefficients A_u through E_1 have been introduced in Chap. 3. The applications of these equations are explored in the following sections.

5.3   Longitudinal Control

5.3.1   Fundamentals

The longitudinal control is the control of any linear/rotational motion control in the x–z plane (e.g., pitch about y-axis, plunging, climbing, cruising, pulling up, and descending). Any change in thrust, lift, drag, and pitching moment have the major influence on this motion. Longitudinal aerodynamic control of an aircraft can be achieved by providing an incremental lift force on horizontal tail or wing. The pitch control is the heart of a longitudinal control.

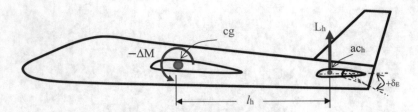

Fig. 5.3 Longitudinal aerodynamic control via elevator deflection

In a fixed-wing conventional aircraft (e.g., Boeing 777 and Cessna Citation), the elevator is movable part of the horizontal tail (aft or canard). The structure of an elevator—located at the tail trailing edge—is very similar to a wing flap, but it is deflected up and down.

The pitching moment in an aircraft is generated by: (1) changing the lift coefficient of the horizontal tail via deflecting the elevator, and (2) changing the throttle setting which consequently varies the engine thrust.

The aerodynamic pitching moment (M) is primarily a function of horizontal tail lift (L_{ht}) and the tail arm. We shall examine how an elevator on a horizontal tail provides the required control moment. By definition, the down deflection of an elevator (see Fig. 5.3) is assumed a positive deflection ($+ \delta_E$) which produces a positive tail lift force ($+ L_h$), while a negative (counter-clockwise) pitching moment ($- \Delta M$) is generated. This moment is equal to the horizontal tail lift multiplied by tail arm (i.e., the distance between tail ac to aircraft cg; l_h).

$$\Delta M = \Delta L_h l_h \tag{5.21}$$

This moment can also be calculated directly from elevator deflection:

$$\Delta M = \frac{1}{2}\rho V^2 S C (\Delta C_m) \tag{5.22}$$

where the change in pitching moment coefficient is:

$$\Delta C_m = C_{m_{\delta_E}} \Delta \delta_E \tag{5.23}$$

The parameter $C_{m_{\delta_E}}$ is called the non-dimensional longitudinal control power derivative or elevator control power derivative. The larger the value of $C_{m_{\delta_E}}$, the more effective the control is in creating the longitudinal control moment. Note that, the parameter $\Delta \delta_E$ is employed, which implies, it is a deflection beyond what is needed for longitudinal trim.

$$\delta_{E_{new}} = \delta_{Eo} + \Delta \delta_E \tag{5.24}$$

where δ_{Eo} is the initial elevator deflection that was used to longitudinally trim the aircraft, before a control desire is applied. The longitudinal control moment produces an angular acceleration about y-axis, or pitch rate:

$$\Delta M = I_{yy}\dot{q} \tag{5.25}$$

This equation is reformatted by using dimensional control derivatives (here, M_{δ_E}):

$$\dot{q} = \ddot{\theta} = M_{\delta_E}\Delta\delta_E \tag{5.26}$$

The parameter M_{δ_E} is a control derivative and is called the dimensional elevator control power derivative, which is obtained by:

$$M_{\delta_E} = \frac{\overline{q}_1 S \overline{C} C_{m_{\delta_E}}}{I_{yy}U_o} \tag{5.27}$$

where non-dimensional derivative $C_{m_{\delta_E}}$ is the rate of change of pitching moment coefficient due to elevator deflection ($\frac{\partial C_m}{\partial \delta_E}$) and is obtained by:

$$C_{m_{\delta_E}} = -C_{L_{\alpha_h}}\eta_h V_H \tau_e \tag{5.28}$$

where τ_e is the angle of attack effectiveness of the elevator (represents the elevator effectiveness) and is determined from Fig. 5.4. The typical value for derivative $C_{m_{\delta_E}}$ is about -0.2 1/rad to -4 1/rad. The elevator longitudinal control power is a function of: (1) Elevator chord, (2) Elevator span, (3) Elevator deflection, (4) Horizontal tail arm, (5) Horizontal tail planform area, (6) Horizontal tail airfoil, and (7) Horizontal tail efficiency. As any of these parameters increased/improved, the elevator will be more effective is pitch control. Moreover, the ability of the elevator to pitch the aircraft (i.e., elevator control power) depends on altitude (i.e., air density) and airspeed.

Example 5.1 Consider a fixed-wing General Aviation aircraft with a maximum takeoff weight of 2000 lb, a wing planform area of 150 ft^2, is cruising with a speed of 90 knot at 10,000 ft where air density is 17.56×10^{-4} slug/ft^3. Other characteristics are:

$$C_{L_{\alpha_h}} = 5.2\frac{1}{rad}; C = 4.3 \text{ ft}; \eta_h = 0.95; l_h = 14 \text{ ft}; S_h = 40 \text{ ft}^2; I_{yy} = 1200 \text{ slug.ft}^2$$

The elevator chord is 20% of the horizontal tail chord. Pilot applies a $-10°$ of elevator deflection. Determine the angular pitch acceleration ($\ddot{\theta}$) as a response of the aircraft to this input.

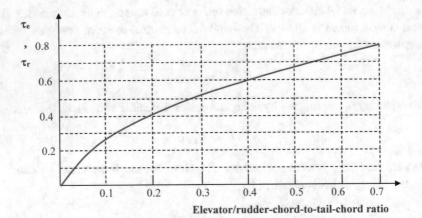

Fig. 5.4 Control surface angle of attack effectiveness parameter

Solution

$$\overline{V}_h = \frac{l_h S_h}{\overline{C} S} = \frac{14 \times 40}{4.3 \times 150} = 0.87 \tag{2.13}$$

From Fig. 5.4, for $C_E/C_h = 0.2$, $\tau_e = 0.41$.

$$C_{m_{\delta_E}} = -C_{L_{\alpha_h}} \eta_h V_H \tau_e = -5.2 \times 0.95 \times 0.87 \times 0.41 = -1.76 \frac{1}{\text{rad}} \tag{5.28}$$

The change in pitching moment coefficient is:

$$\Delta C_m = C_{m_{\delta_E}} \Delta \delta_E = -1.76 \times \frac{-10}{57.3} = 0.307 \tag{5.23}$$

The pitching moment:

$$\Delta M = \frac{1}{2}\rho V^2 SC(\Delta C_m) = \frac{1}{2} \times 0.001756 \times (90 \times 1.688)^2 \times 150 \times 4.3 \times 0.307 \tag{5.22}$$

$$\Delta M = 4010.5 \, \text{lbf.ft}$$

This is a positive (nose-up) pitching moment, as we expected due to negative (i.e., up) elevator deflection.

$$\Delta M = I_{yy}\dot{q} => \dot{q} = \ddot{\theta} = \frac{\Delta M}{I_{yy}} = \frac{4010.5}{1200} = 3.34 \frac{\text{rad}}{\text{s}^2} = 191.5 \frac{\text{deg}}{\text{s}^2} \tag{5.25}$$

5.3.2 Elevator Effectiveness

When the elevator is deflected, it changes two aerodynamic forces and one aerodynamic moment: (1) Aircraft total lift, (2) Horizontal tail lift, and (3) Aircraft pitching moment. The new values of these forces and moment will cause the aircraft to move a new longitudinal trim point. The rate of transition from the initial longitudinal trim point to the new trim point (i.e., pitch rate) is a function of elevator effectiveness.

The aircraft new longitudinal trim point features: (1) a new angle of attack, (2) a new pitch angle, (3) a new climb angle, (4) a new rate of climb/descent, and (5) a new airspeed. Due to these new flight parameters, the aircraft altitude shall continuously vary.

To examine how an elevator provides the required longitudinal control moment, we can evaluate three non-dimensional control derivatives: (1) $C_{m_{\delta_E}}$, (2) $C_{L_{\delta_E}}$, and (3) $C_{L_{h_{\delta_E}}}$. The significance of the derivative $C_{m_{\delta_E}}$—which represents the rate of change of pitching moment coefficient due to elevator deflection—is discussed earlier in this section. The representation of elevator effectiveness by other two derivatives are discussed in this section.

The elevator has the following main geometric characteristics: (1) Elevator-chord-to-tail-chord ratio (C_E/C_h), (2) Elevator-span-to-tail-span ratio (b_E/b_h), (3) Maximum up deflection ($-\delta_{E_{max}}$), and (4) Maximum down deflection ($+\delta_{E_{max}}$). The elevator span is often selected to be equal to the horizontal tail span (i.e., bE/bh = 1). The elevator chord ratio is often selected to be either about 20–30% (for GA and transport aircraft), or equal to 1 (for fighters). In fighter aircraft, an all moving horizontal tail ($C_E = C_h$) is utilized to satisfy high maneuverability requirements. Table 5.1 [3] shows specifications of elevators for several aircraft.

Another measure of elevator effectiveness is a parameter which represents the contribution of elevator to the aircraft lift ($C_{L_{\delta_E}}$). This non-dimensional derivative is the rate of change of aircraft lift coefficient with respect to elevator deflection and is defined as follows:

$$C_{L_{\delta_E}} = \frac{\partial C_L}{\partial \delta_E} = C_{L_{\alpha_h}} \eta_h \frac{S_h}{S} \frac{b_E}{b_h} \tau_e \tag{5.29}$$

The third measure of elevator effectiveness is a non-dimensional derivative which represents the contribution of elevator to the horizontal tail lift ($C_{L_{h_{\delta_E}}}$). This derivative is the rate of change of tail lift coefficient (C_{L_h}) with respect to elevator deflection and is defined as follows:

$$C_{L_{h_{\delta_E}}} = \frac{\partial C_{L_h}}{\partial \delta_E} = \frac{\partial C_{L_h}}{\partial \alpha_h} \frac{\partial \alpha_h}{\partial \delta_E} = C_{L_{\alpha_h}} \tau_e \tag{5.30}$$

When the elevator is deflected, the aircraft begins to pitch with an initial rate of q_o until the aircraft reaches a new longitudinal trim point. This implies that the aircraft will gain a new pitch angle, a new angle of attack, and a new airspeed. During a longitudinal

Table 5.1 Specifications of elevators for several aircraft

No.	Aircraft	Type	m_{TO} (kg)	S_E/S_h	C_E/C_h	δ_{Emax} (deg) Down	Up
1	Cessna 182	Light GA	1406	0.38	0.44	22	– 25
2	Cessna Citation III	Business jet	9979	0.37	0.37	15	– 15.5
3	Gulfstream 200	Business jet	16,080	0.28	0.31	20	– 27.5
4	Lockheed C-130 Hercules	Military cargo	70,305	0.232	0.35	15	– 40
5	Boeing 737-100	Transport	50,300	0.224	0.25	20	– 20
6	Boeing 747-200	Transport	377,842	0.185	0.23	17	– 22
7	Boeing 777-200	Transport	247,200	0.30	0.32	25	– 30
8	Airbus 320	Transport	78,000	0.31	0.32	17	– 30
9	Airbus A340-600	Transport	368,000	0.24	0.31	15	– 30
10	Lockheed C-5A	Military cargo	381,000	0.268	0.35	10	– 20
11	General dynamics F-16 fighting falcon	Fighter	19,200	1	1	25	– 25
12	Eurofighter typhoon	Fighter	23,500	1[a]	1	30	– 30

[a] Canard and elevon

control process, the angular pitch acceleration ($\ddot{\theta}$) and pitch rate (q) are also not constant. They both begin with their highest values and then logarithmically approach to zero.

When C_m (in Eq. 5.14) is not zero; it implies that, the elevator has created a pitching moment for longitudinal control. When C_m becomes zero, the aircraft has gone to a new trim point (i.e., it is at longitudinal trim).

Example 5.2 Consider a GA aircraft with a weight of 2400 lb and a wing area of 170 ft^2 and the following stability and control derivatives:

$$X_\alpha = 28.3\frac{ft}{s^2}; \ X_u = -0.036\frac{1}{s}; \ X_{\delta_E} = -0.693\frac{ft}{s^2}$$

$$M_\alpha = -31.57\frac{1}{s^2}; \ M_u = 0.0002\frac{1}{ft.s}; \ M_q = -3.75\frac{1}{s}; \ M_{\dot{\alpha}} = -1.414\frac{1}{s}; \ M_{\delta_E} = -21\frac{1}{s^2}$$

$$Z_\alpha = -392.7\frac{ft}{s^2}; \ Z_u = -0.386\frac{1}{s}; \ Z_q = -9.17\frac{ft}{s}; \ Z_{\dot{\alpha}} = -1.74\frac{ft}{s}; \ Z_{\delta_E} = -26.96\frac{ft}{s^2}$$

The aircraft is cruising with an airspeed of 100 knot and an angle of attack of 2° at sea level. The pilot applies −2° of elevator deflection. A. Plot the variations of angle of attack, airspeed, and pitch angle as the aircraft response to this input. B. Describe the aircraft new longitudinal trim point.

Solution

To determine the aircraft response to the elevator input, first, transfer functions are determined. The coefficients of fourth-order denominator of transfer functions:

$$A_1 = U_o - Z_{\dot{\alpha}} = 170.5 \frac{\text{ft}}{\text{s}}$$

$$B_1 = -(U_o - Z_{\dot{\alpha}})(X_u + M_q) - Z_\alpha - M_{\dot{\alpha}}(U_o + Z_q) = 1266.9 \frac{\text{ft}}{\text{s}^2}$$

$$C_1 = X_u\{M_q(U_o - Z_{\dot{\alpha}}) + Z_\alpha + M_{\dot{\alpha}}(U_o + Z_q)\} + M_q Z_\alpha - Z_u X_\alpha$$
$$+ M_{\dot{\alpha}} g \sin \Theta_o - M_\alpha(U_o + Z_q) = 6588 \frac{\text{ft}}{\text{s}^3}$$

$$D_1 = g \sin \Theta_o(M_\alpha - M_{\dot{\alpha}} X_u) + g \cos \Theta_o\{Z_u M_{\dot{\alpha}} + M_u(U_o - Z_{\dot{\alpha}})\}$$
$$- M_u X_\alpha(U_o + Z_q) + Z_u X_\alpha M_q + X_u\{M_\alpha(U_o + Z_q) - M_q Z_\alpha\} = 257 \frac{\text{ft}}{\text{s}^4}$$

$$E_1 = g \cos \Theta_o\{M_\alpha Z_u - Z_\alpha M_u\} + g \sin \Theta_o(M_u X_\alpha - X_u M_\alpha) = 393 \frac{\text{ft}}{\text{s}^5} \quad (3.38)$$

The coefficients for numerator of airspeed-to-elevator-deflection transfer function are:

$$A_u = X_{\delta_E}(U_o - Z_{\dot{\alpha}}) = -118.2 \frac{\text{ft}^2}{\text{s}^3}$$

$$B_u = -X_{\delta_E}\{M_q(U_o - Z_{\dot{\alpha}}) + Z_\alpha + M_{\dot{\alpha}}(U_o + Z_q)\} + Z_{\delta_E} X_\alpha = -1635.6 \frac{\text{ft}^2}{\text{s}^4}$$

$$C_u = X_{\delta_E}\{M_q Z_\alpha + M_{\dot{\alpha}} g \sin \Theta_o - M_{\dot{\alpha}}(U_o + Z_q)\} + Z_{\delta_E}\{-M_{\dot{\alpha}} g \cos \Theta_o - X_\alpha M_q\}$$
$$+ M_{\delta_E}\{X_\alpha(U_o + Z_q) - (U_o - Z_{\dot{\alpha}})g \cos \Theta_o\} = 11{,}655 \frac{\text{ft}^2}{\text{s}^5}$$

$$D_u = X_{\delta_E} M_\alpha g \sin \Theta_o - Z_{\delta_E} M_\alpha g \cos \Theta_o$$
$$+ M_{\delta_E}(Z_\alpha g \cos \Theta_o - X_\alpha g \sin \Theta_o) = 239{,}092 \frac{\text{ft}^2}{\text{s}^6} \quad (3.39)$$

The coefficients for numerator of angle-of-attack-to-elevator-deflection transfer function are:

$$A_\alpha = Z_{\delta_E} = -26.96 \frac{\text{ft}}{\text{s}^2}$$

$$B_\alpha = X_{\delta_E} Z_u - Z_{\delta_E}(M_q + X_u) + M_{\delta_E}(U_o + Z_q) = -3461.5 \frac{\text{ft}}{\text{s}^3}$$

$$C_\alpha = X_{\delta_E}\{(U_o + Z_q)M_u - M_q Z_u\} + Z_{\delta_E} M_q X_u$$
$$- M_{\delta_E}\{g \sin \Theta_o + X_u(U_o + Z_q)\} = -160.2 \frac{\text{ft}}{\text{s}^4}$$

$$D_\alpha = -X_{\delta_E} M_u g \sin \Theta_o + Z_{\delta_E} M_u g \cos \Theta_o$$
$$+ M_{\delta_E}(X_u g \sin \Theta_o - Z_u g \cos \Theta_o) = -260 \frac{\text{ft}}{\text{s}^5} \tag{3.40}$$

The coefficients for numerator of pitch-angle-to-elevator-deflection transfer function are:

$$A_\theta = Z_{\delta_E} M_{\dot\alpha} + M_{\delta_E}(U_o - Z_{\dot\alpha}) = -3551 \frac{\text{ft}}{\text{s}^3}$$

$$B_\theta = X_{\delta_E}\{Z_u M_{\dot\alpha} + M_u(U_o - Z_{\dot\alpha})\} + Z_{\delta_E}(M_\alpha - M_{\dot\alpha} X_u)$$
$$- M_{\delta_E}\{Z_\alpha + X_u(U_o - Z_{\dot\alpha})\} = -7541 \frac{\text{ft}}{\text{s}^4}$$

$$C_\theta = X_{\delta_E}(M_\alpha Z_u - Z_\alpha M_u) + Z_{\delta_E}(X_\alpha M_u - M_\alpha X_u)$$
$$+ M_{\delta_E}(Z_\alpha X_u - X_\alpha Z_u) = -505 \frac{\text{ft}}{\text{s}^5} \tag{3.41}$$

Thus, transfer functions are:

$$\frac{u(s)}{\delta_E(s)} = \frac{A_u s^3 + B_u s^2 + C_u s + D_u}{A_1 s^4 + B_1 s^3 + C_1 s^2 + D_1 s + E_1}$$
$$= \frac{-118.2 s^3 - 1635.6 s^2 + 11655 s + 239,092}{170.5 s^4 + 1266.9 s^3 + 6588 s^2 + 257 s + 393} \tag{5.17}$$

$$\frac{\alpha(s)}{\delta_E(s)} = \frac{A_\alpha s^3 + B_\alpha s^2 + C_\alpha s + D_\alpha}{A_1 s^4 + B_1 s^3 + C_1 s^2 + D_1 s + E_1}$$
$$= \frac{-26.96 s^3 - 3461.5 s^2 - 160.2 s - 260}{170.5 s^4 + 1266.9 s^3 + 6588 s^2 + 257 s + 393} \tag{5.18}$$

$$\frac{\theta(s)}{\delta_E(s)} = \frac{A_\theta s^2 + B_\theta s + C_\theta}{A_1 s^4 + B_1 s^3 + C_1 s^2 + D_1 s + E_1}$$
$$= \frac{-3551 s^2 - 7541 s - 505}{170.5 s^4 + 1266.9 s^3 + 6588 s^2 + 257 s + 393} \tag{5.19}$$

To simulate the flight, the following matlab code is written:

```
t = 0:0.1:400; %sec
dE=(-2/57.3)+t*0; % rad
G_u=tf([-118.2 -1635.6 11654.6 239092],[170.5 1267 6588 257 393]);
```

```
G_a=tf([-27 -3461.5 -160.2 -260],[170.5 1267 6588 257 393]);
G_t=tf([-3551 -7541.4 -504.8],[170.5 1267 6588 257 393]);
y1= lsim(G_a,dE,t);
y2=lsim(G_t,dE,t);
y3=lsim(G_u,dE,t);
subplot 311
plot(t,y1*57.3+2,'*r')
ylabel('\alpha (deg)')
grid
subplot 312
plot(t,y2*57.3+2,'og')
ylabel('\theta (deg)')
grid
subplot 313
plot(t,(y3+169)/1.688,'+b')
ylabel('speed (knot)')
xlabel('time(sec)')
grid
```

When the program is executed in matlab, the plots in Fig. 5.5 are generated. Theses plots indicate that the aircraft have these new flight parameters:

$$\alpha_{new} = 3.3 \, \text{deg}, \theta_{new} = 5 \, \text{deg}, u_{new} = 87 \, \text{knot}$$

Thus, the aircraft continue climbing with a climb angle of $1.7°$ ($5-3.3 = 1.7$). The aircraft will reach this new trim condition after about 400 s. Note that, the engine thrust remains the same and was not increased.

5.4 Takeoff Rotation Control

There are various flight conditions that require a specific motion or aerodynamic force and/or pitching moment from the elevator. For a conventional fixed-wing aircraft with a landing gear configuration which the main gear is behind aircraft cg (e.g., tricycle landing gear), the take-off rotation (about the main landing gear) is the most challenging longitudinal control task. During a takeoff, the elevator control power must be sufficiently large to rotate the aircraft (i.e., lift the nose wheel off the ground), while the main gear is still on the runway. During takeoff which the aircraft is flying very slowly, it is necessary that enough elevator effectiveness be available to rotate the fuselage nose with a desired pitch rate. In this Section, the analysis of takeoff rotation control is presented.

According to the longitudinal control handling qualities requirements for take-off; in an aircraft with a tricycle landing gear, the pitch rate (q) should have a value such that the take-off rotation does not take longer than a specified length of time (e.g., a few seconds).

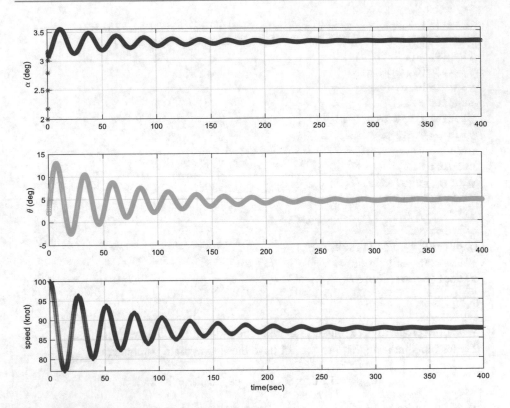

Fig. 5.5 Pitch response to − 2° of elevator deflection

In large transport aircraft, the correct takeoff attitude is achieved in approximately 3–4 s after rotation.

This requirement must be satisfied when the cg is at its most forward position. On the Boeing 777–200, the cg positions from 14 to 44% of MAC are allowed. Thus, the most forward cg position of a Boeing 777–200 is at 14% of MAC.

A desired angular acceleration about the main gear rotation point; $\ddot{\theta}$ is the primary longitudinal control requirement. Typical rotational acceleration is given in Table 5.2 for various types of fixed-wing aircraft. In transport aircraft Boeing 737, it is recommended to raise the aircraft's nose smoothly at a rate of no more than 2–3° per second toward an initial target pitch attitude of 8–10° (15° maximum). The correct takeoff attitude is achieved in approximately 3–4 s after rotation.

The aircraft speed at the time of takeoff rotation (V_R) is often slightly greater than the stall speed (V_s):

$$V_R = k_R V_s \qquad (5.31)$$

Table 5.2 Take-off angular acceleration requirement

No.	Aircraft type	Rotation time during take-off (s)	Take-off pitch angular acceleration (deg/s^2)
1	Highly maneuverable (e.g., fighter and acrobatic)	0.2–0.7	12–20
2	Utility; semi-acrobatic GA	1–2	10–15
3	Normal general aviation	1–3	8–10
4	Small transport	2–4	6–8
5	Large transport	3–5	4–6
6	Civil unmanned aircraft	1–2	10–15

where $k_R = 1.1$–1.3. To include a safety factor, the elevator is designed to rotate the aircraft with the desired acceleration at the stall speed (V_s).

Consider a fixed-wing aircraft with a tricycle landing gear in Fig. 5.6 which is at the onset of a rotation about the main gear in a take-off operation. The figure illustrates all forces and moments contributing to this moment of the take-off operation. Contributing forces include wing-fuselage lift (L_{wf}), horizontal tail lift (L_h), aircraft drag (D), friction force between tires and the ground (F_f), aircraft weight (W), engines thrust (T), and inertia (acceleration) force (m.a).

The latter force (m.a) is acting backward due to the Newton's third law as a reaction to the acceleration. Furthermore, the contributing moments are the wing-fuselage aerodynamic pitching moment (Mo_{wf}) plus the moments of preceding forces about the rotation point (i.e., main gear). The distances between these forces are measured with respect to

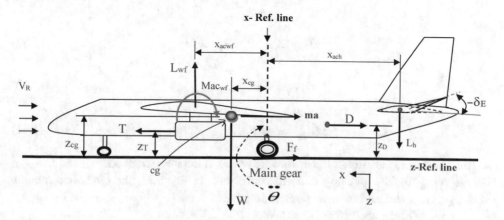

Fig. 5.6 Forces and moments during take-off rotation

both x-reference line (i.e., main gear), and z-reference line (i.e., ground). The ground friction coefficient, μ, depends on the type of terrain. The wheel-ground friction coefficient is about 0.02–0.04 for concrete and asphalt runways; and 0.1–0.3 for a soft ground.

There are three governing equations of that govern the aircraft motion at the instant of rotation—two force equations and one moment equation:

$$\sum F_x = m\frac{dV}{dt} \Rightarrow T - D - F_f = ma \Rightarrow T - D - \mu N = ma \qquad (5.32)$$

$$\sum F_z = 0 \Rightarrow L + N = W \Rightarrow L_{wf} - L_h + N = W \Rightarrow N = W - \left(L_{wf} - L_h\right) \quad (5.33)$$

$$\sum M_{cg} = I_{yy_{mg}}\ddot{\theta} \Rightarrow -M_W + M_D - M_T + M_{L_{wf}}$$
$$+ M_{ac_{wf}} + M_{L_h} + M_a = I_{yy_{mg}}\ddot{\theta} \qquad (5.34)$$

The normal force (N), the friction force (F$_f$), and the aircraft lift at take-off are:

$$N = W - L_{TO} \qquad (5.35)$$

$$F_f = \mu N = \mu(W - L_{TO}) \qquad (5.36)$$

$$L_{TO} = L_{wf} + L_h \qquad (5.37)$$

The wing-fuselage lift (L$_{wf}$), horizontal tail lift (L$_h$), aerodynamic drag (D) forces and wing-fuselage pitching moment about wing-fuselage aerodynamic center are shown below. Recall that the horizontal tail lift is negative.

$$L_h = \frac{1}{2}\rho V_R^2 C_{L_h} S_h \qquad (5.38)$$

$$L_{wf} \cong \frac{1}{2}\rho V_R^2 C_{L_{TO}} S_{ref} \qquad (5.39)$$

$$D_{TO} = \frac{1}{2}\rho V_R^2 C_{D_{TO}} S_{ref} \qquad (5.40)$$

$$M_{ac_{wf}} = \frac{1}{2}\rho V_R^2 C_{m_{acwf}} S_{ref}\overline{C} \qquad (5.41)$$

where V$_R$ denote the aircraft linear forward speed at the instant or rotation, and Cm$_{ac\_wf}$ denote wing-fuselage pitching moment coefficient. In Eq. 5.31, the clockwise rotation about y-axis is assumed to be as positive rotation.

The aircraft drag coefficient [4] at takeoff (C$_{D_{oTO}}$) is:

$$C_{D_{TO}} = C_{D_{oTO}} + KC_{L_{TO}}^2 \qquad (5.42)$$

where K is the induced drag factor and is:

$$K = \frac{1}{\pi.e.AR} \tag{5.43}$$

The aircraft lift coefficient at takeoff ($C_{L_{TO}}$) is a function of cruise lift coefficient (C_{L_C}) and extra lift coefficient generated by flap ($\Delta C_{L_{flap}}$):

$$C_{L_{TO}} = C_{L_C} + \Delta C_{L_{flap}} \tag{5.44}$$

The contributing pitching moments in take-off rotation control are aircraft weight moment (M_W), aircraft drag moment (M_D), engine thrust moment (M_T), wing-fuselage lift moment (M_{Lwf}), wing-fuselage aerodynamic pitching moment ($M_{ac_{wf}}$), horizontal tail lift moment (M_{Lh}), and linear acceleration moment (M_a). These longitudinal moments are obtained as follows:

$$M_W = W(x_{cg}) \tag{5.45}$$

$$M_D = D(z_D) \tag{5.46}$$

$$M_T = T(z_T) \tag{5.47}$$

$$M_{Lwf} = L_{wf}(x_{ac_{wf}}) \tag{5.48}$$

$$M_{Lh} = L_h(x_{ac_h}) \tag{5.49}$$

$$M_a = ma(z_{cg}) \tag{5.50}$$

The inclusion of the moment generated by the aircraft acceleration (Eq. 5.47) is due to the fact that based on the Newton's third law; any action creates a reaction (m.a). This reaction force is producing a moment when its corresponding arm is taken into account. Substituting these moments into Eq. (5.31) yields:

$$-W(x_{cg}) + D(z_D) - T(z_T) + L_{wf}(x_{ac_{wf}}) + M_{ac_{wf}} - L_h(x_{ac_h}) + ma(z_{cg}) = I_{yy_{mg}}\ddot{\theta} \tag{5.51}$$

where $I_{yy_{mg}}$ represents the aircraft mass moment of inertia about y-axis at the main gear. In an aircraft with a tricycle landing gear, the tail lift moment, wing-fuselage moment, drag moment, and acceleration moment are all clockwise, while the weight moment, thrust moment, and wing-fuselage aerodynamic pitching moment are counterclockwise. These directions must be considered when assigning a sign to each term. The role of elevator in Eq. 5.51 is to create a sufficient horizontal tail lift (L_h). The result is as follows:

$$L_h = \frac{L_{wf}(x_{ac_{wf}}) + M_{ac_{wf}} + ma(z_{cg}) - W(x_{cg}) + D(z_D) - T(z_T) - I_{yy_{mg}}\ddot{\theta}}{x_{ac_h} - x_{mg}} \tag{5.52}$$

To find represents the aircraft mass moment of inertia about y-axis at the main gear (Iyy_{mg}), use parallel axis theorem:

$$I_{yy_{mg}} = I_{yy} + md^2 \tag{5.53}$$

where d is the distance between aircraft cg and the point of contact between main gear and the ground:

$$d = \sqrt{x_{cg}^2 + z_{cg}^2} \tag{5.54}$$

The horizontal tail lift must be such that to satisfy the take-off rotation requirement. The elevator contribution to this lift is through tail lift coefficient which can be obtained by:

$$C_{L_h} = \frac{2L_h}{\rho V_R^2 S_h} \tag{5.55}$$

This maximum tail lift coefficient is generally negative (about -1 to -1.5) and is a function of tail angle of attack (α_h), tail airfoil section, tail geometry and elevator deflection. The horizontal tail lift coefficient is linearly modeled as:

$$C_{L_h} = C_{L_{ho}} + C_{L_{\alpha_h}}\alpha_h + C_{L_{h\delta E}}\delta_E \tag{5.56}$$

where $C_{L_{\alpha_h}}$ is the tail lift curve slope and $C_{L_{ho}}$ is the zero angle of attack tail lift coefficient. Most horizontal tails tend to use a symmetric airfoil section, so the parameter $C_{L_{ho}}$ is normally zero. Thus:

$$C_{L_h} = C_{L_{\alpha_h}}\alpha_h + C_{L_{\alpha_h}}\tau_e\delta_E = C_{L_{\alpha_h}}(\alpha_h + \tau_e\delta_E) \tag{5.57}$$

The tail angle of attack is defined as:

$$\alpha_h = \alpha + i_h - \varepsilon \tag{5.58}$$

where α is the aircraft angle of attack at the onset of rotation, i_h denotes the tail incidence angle, and ε represents the downwash angle. The aircraft angle of attack, when the aircraft is on the ground (i.e., onset of rotation) is usually zero.

Example 5.3 Figure 5.7 illustrates the geometry of a high-wing twin jet engine light utility aircraft which is equipped with a tricycle landing gear. The aircraft has the following characteristics.

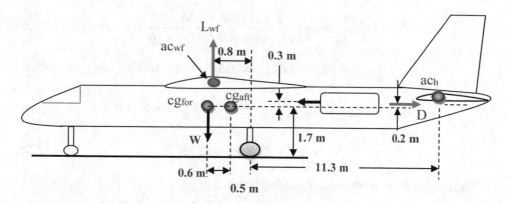

Fig. 5.7 Geometry of aircraft for Example 5.3

$m_{TO} = 20{,}000$ kg, $V_R = 85$ knot, $I_{yy} = 150{,}000$ kg m^2, $T_{max} = 2 \times 28$ kN, $L_f = 23$ m, $V_C = 360$ knot (at 25,000 ft), $C_{L_o} = 0.24$, $C_{D_{oC}} = 0.024$, $C_{D_{oTO}} = 0.038$.

Wing:

$$S = 70\,\text{m}^2, \text{AR} = 8, \text{e} = 0.88, \lambda = 1, \Delta C_{L_{flap}} = 0.5,$$

$$C_{mac_{wf}} = 0.05, i_w = 2\,\text{deg}, h_o = 0.25$$

Horizontal tail:
$S_h = 16\,\text{m}^2$, $b_h = 9\,\text{m}$, $C_{L_{\alpha_h}} = 4.31/\text{rad}$, $i_h = -1\,\text{deg}$, $\lambda_h = 1$,

$\eta_h = 0.96$, Airfoil: NACA 0009, $\varepsilon = 1.5\,\text{deg}$, $C_E/C_h = 0.3$
How much elevator must be deflected at the take-off rotation to create a pitch angular acceleration of 12 deg/s^2?

Solution

The air density at sea level is 1.225 kg/m^3, and at 25,000 ft is 0.549 kg/m^3. To obtain the wing mean aerodynamic chord, the following calculations are made:

$$b = \sqrt{S.\text{AR}} = \sqrt{70 \times 8} = 23.66\,\text{m} \tag{3.65}$$

$$\overline{C} = \frac{S}{b} = \frac{70}{23.66} = 2.96\,\text{m} \tag{3.66}$$

To find the aircraft drag, we have:

$$K = \frac{1}{\pi.\text{e.AR}} = \frac{1}{3.14 \times 0.88 \times 8} = 0.045 \tag{5.43}$$

$$C_{LC} = \frac{2W}{\rho V_C^2 S} = \frac{2 \times 20,000 \times 9.81}{0.549 \times (360 \times 0.5144)^2 \times 70} = 0.297 \tag{2.27}$$

$$C_{L_{TO}} = C_{LC} + \Delta C_{L_{flap}} = 0.297 + 0.5 = 0.797 \tag{5.44}$$

$$C_{D_{TO}} = C_{D_oTO} + KC_{L_{TO}}^2 = 0.038 + 0.045 \times 0.797^2 = 0.067 \tag{5.42}$$

$$V_R = V_S = 85 \text{ knot} = 43.73 \frac{m}{s} \tag{5.31}$$

The longitudinal aerodynamic forces and moment:

$$D_{TO} = \frac{1}{2}\rho_o V_R^2 SC_{D_{TO}} = \frac{1}{2} \times 1.225 \times (43.73)^2 \times 70 \times 0.067 = 5472 \text{ N} \tag{5.40}$$

$$L_{TO} \approx L_{wf} = \frac{1}{2}\rho_o V_R^2 S_{ref} C_{L_{TOf}}$$
$$= \frac{1}{2} \times 1.225 \times (43.73)^2 \times 70 \times 0.797 = 65,371 \text{ N} \tag{5.37}$$

$$M_{ac_{wf}} = \frac{1}{2}\rho_o V_R^2 C_{mac_{wf}} S_{ref} \bar{C}$$
$$= \frac{1}{2} \times 1.225 \times (43.73)^2 \times (0.05) \times 70 \times 2.96 = 12,125 \text{ Nm}$$

The runway is assumed to be concrete, so a ground-friction coefficient of 0.04 is selected.

$$F_f = \mu(W - L_{TO}) = 0.04(20,000 \times 9.81 - 65,371) = 5230.5 \text{ N} \tag{5.36}$$

Aircraft linear acceleration at the time of take-off rotation

$$a = \frac{T - D_{TO} - F_R}{m} = \frac{2 \times 28,000 - 5472 - 5230.5}{20,000} \Rightarrow a = 2.265 \frac{m}{s^2} \tag{5.32}$$

The clockwise rotation about y axis is considered to be as positive direction. The most forward cg is the most critical case.

$$M_W = W(x_{mg} - x_{cg}) = -20,000 \times 9.81 \times 1.1 = -215,746 \text{ Nm} \tag{5.45}$$

$$M_D = D(z_D - z_{mg}) = 5472 \times 1.9 = 10,397 \text{ Nm} \tag{5.46}$$

$$M_T = T(z_T - z_{mg}) = -2 \times 28,000 \times (1.7 + 0.3) = -112,000 \text{ Nm} \tag{5.47}$$

$$M_{L_{wf}} = L_{wf}(x_{mg} - x_{ac_{wf}}) = 65{,}371 \times 0.8 = 52{,}297 \, \text{Nm} \tag{5.48}$$

$$M_a = ma(z_{cg} - z_{mg}) = 20{,}000 \times 2.265 \times 1.7 = 77{,}005.5 \, \text{Nm} \tag{5.50}$$

Horizontal tail lift:

$$L_h = \cfrac{\begin{array}{c} L_{wf}(x_{mg} - x_{ac_{wf}}) + M_{ac_{wf}} + ma(z_{cg} - z_{mg}) + W(x_{mg} - x_{cg}) \\ + D(z_D - z_{mg}) + T(z_T - z_{mg}) - I_{yy_{mg}}\ddot{\theta} \end{array}}{x_{ac_h} - x_{mg}} \tag{5.52}$$

$$L_h = \cfrac{\begin{array}{c} 52{,}297 + 12{,}125 + 77{,}005.5 - 215{,}746 + 10{,}397 \\ - 112{,}000 - \left(150{,}000 \times \dfrac{12}{57.3}\right) \end{array}}{11.3} = -18{,}348 \, \text{N}$$

The horizontal tail lift coefficient:

$$C_{L_h} = \frac{2L_h}{\rho_o V_R^2 S_h} = \frac{2 \times (-18{,}348)}{1.225 \times 43.73^2 \times 16} \Rightarrow C_{L_h} = -0.979 \tag{5.55}$$

Elevator deflection:
During take-off, the fuselage is horizontal, so $\alpha = 0$.

$$\alpha_h = \alpha + i_h - \varepsilon = 0 - 1 + 1.5 = 0.5 \, \text{deg} \tag{5.58}$$

Tail has a symmetric airfoil, $C_{L_{ho}} = 0$. Since the elevator to tail chord ratio is 30%, from Fig. 5.4, $\tau_e = 0.52$.

$$C_{L_h} = C_{L_{\alpha_h}}(\alpha_h + \tau_e \delta_E) \tag{5.57}$$

$$\delta_E = \frac{1}{\tau_e}\left[\frac{C_{L_h}}{C_{L_{\alpha_h}}} - \alpha_h\right] = \frac{1}{0.52}\left[\frac{-0.979}{4.3} - \frac{0.5}{57.3}\right] = -0.45 \, \text{rad} = -26 \, \text{deg}$$

5.5 Pure Pitching Motion Approximation

In a pure pitching motion, the sum of all pitching moments (M) will be applied about aircraft cg to change the pitch angle (θ). By assuming "the change in α and are identical", one can derive the following approximate governing differential equation for longitudinal motion:

$$\ddot{\alpha} - (M_q + M_{\dot{\alpha}})\dot{\alpha} - M_\alpha \alpha = M_{\delta_E}\delta_E \tag{5.59}$$

By applying LaPlace transform, this differential equation is readily converted to a transfer function:

$$s^2\alpha(s) + \left[-(M_q + M_{\dot{\alpha}})\right]s\alpha(s) - M_\alpha\alpha(s) = M_{\delta_E}.\delta_E(s) \quad (5.60)$$

or

$$\alpha(s)\left[s^2 - s(M_q + M_{\dot{\alpha}}) - M_\alpha\right] = M_{\delta_E}.\delta_E(s) \quad (5.61)$$

Thus, the transfer function is obtained as:

$$\frac{\alpha(s)}{\delta_E(s)} = \frac{M_{\delta_E}}{s^2 - s(M_q + M_{\dot{\alpha}}) - M_\alpha} \quad (5.62)$$

This is an approximate for the angle-of-attack-to-elevator-deflection transfer function.

5.6   Stick Force and Hinge Moment

In a manned aircraft, control surfaces (e.g., elevator) are controlled by the body force of a human pilot. The pilot's link to the elevator/aileron is commonly by use of the stick/yoke/wheel, and the pilot's link to the rudder is commonly by use of the pedals. One of the objectives of aerodynamic design of control surfaces and their power transmission mechanisms is to reduce the hinge moment and the required force felt by pilot. In designing an elevator, great care must be used, so that the stick (control column) force is within the acceptable limits for the pilots. In this section, the stick force, and its related topics; hinge moment and trim tab are discussed.

5.6.1   Hinge Moment

In order to ensure that the pilot is fully and comfortably capable of moving the stick/pedal and deflect the control surfaces, the aerodynamic force and hinge moment of a control surface must be balanced or reduced. The stick force at the cockpit control column is a function of elevator hinge moment and the power transmission mechanism.

The hinge moment created by a control surface must be such that the pilot is comfortably capable of handling the moment. Furthermore, the effort should be small enough to ensure that the pilot does not fatigue in a prolong application (e.g., a few hours).

The pilot force applied to a stick/yoke/wheel/pedal is transmitted to a control surface via a power transmission system which may be mechanical, hydraulic, and electric. For a small normal GA aircraft, a simple mechanical mechanism to directly drive elevator is employed (see Fig. 5.8). The elevator is rotated by the pilot via stick/yoke. When the pilot

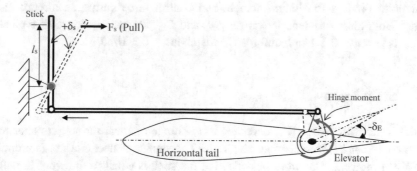

Fig. 5.8 Stick-elevator mechanical mechanism

pulls the stick/yoke; the elevator is deflected upward; whereas; when the pilot pushes the stick/yoke; the elevators is deflected downward.

The hinge moment created by a control surface (here, elevator) is modeled similar to other aircraft aerodynamic moments as:

$$H = \frac{1}{2}\rho U_1^2 S_E C_E C_h \tag{5.63}$$

where S_E denotes the planform area of the elevator, and C_E denotes the mean aerodynamic chord of the elevator. The parameter C_h is the hinge moment coefficient and is given by:

$$C_h = C_{h_o} + C_{h_{\alpha_h}}\alpha_h + C_{h_{\delta_E}}\delta_E \tag{5.64}$$

where α_h is the angle of attack of the horizontal tail, and δ_t is the tab deflection. The parameter C_{ho} is the hinge moment coefficient for $\alpha_c = \delta_c = 0$, and is zero for a symmetrical airfoil. The parameters C_{h_α} and $C_{h_{\delta_c}}$ are two non-dimensional derivatives as:

$$C_{h_\alpha} = \frac{\partial C_h}{\partial \alpha_{LS}} \tag{5.65}$$

$$C_{h_{\delta_C}} = \frac{\partial C_h}{\partial \delta_C} \tag{5.66}$$

These two derivatives are the partial derivatives of hinge moment coefficient (C_h) with respect to lifting surface angle of attack (α_{LS}) and control surface deflection (δ_c) respectively. The hinge moment derivative; C_{h_δ} sometimes referred to as the control heaviness parameter; is the main parameter to reduce the hinge moment and control force. This derivative is a function of a number of variables including the distance between ac of the control surface to the hinge, and the control surface nose curvature. The hinge moment derivative; $C_{h_{\alpha_h}}$ sometimes referred to as the control floating parameter; is the main parameter to produce the hinge moment and control force during an aircraft response to a gust. Both hinge moment derivatives C_{h_α} and $C_{h_{\delta_c}}$ are usually negative; typical value for $C_{h_{\alpha_h}}$ is about -0.1 1/rad; and for $C_{h_{\delta_c}}$ is about -0.3 1/rad.

5.6.2 Stick Force

The stick/yoke/wheel force (F_s) is related to the hinge moment through a factor referred to as power transmission system gearing ratio. The stick to deflect elevator is proportional to the hinge moment. The work of deflecting the stick is equal to the work in deflecting the elevator:

$$F_s l_s \delta_s = H \delta_e \tag{5.67}$$

This can be reduced to:

$$F_s = G_e H \tag{5.68}$$

where G_e is the ratio between the linear/angular movement of the stick/wheel to deflection of the control surface (referred to as the stick to elevator gearing ratio).

$$G_e = \frac{\delta_e}{l_s \delta_s} \tag{5.69}$$

where l_s is the length of the stick, and δ_s is the stick rotation angle. The elevator gearing ratio is the ratio of the elevator deflection to the stick movement. By inserting hinge moment (Eq. 5.63) into Eq. 5.68, we obtain:

$$F_s = \frac{1}{2} \rho V^2 S_E C_E C_h G_e \tag{5.70}$$

From this expression we see that the magnitude of the stick force increases with the size of the elevator, air density, and the square of the airspeed.

Obtaining linearly varying pilot forces for all flight conditions is an important handling quality requirement for both military and civil aircraft. There are various specific requirements in the FAR regulations for the maximum stick forces and stick force per g.

The stick force created by a control surface must be such that the pilot is comfortably capable of handling the force. Furthermore, the force should be small enough to ensure that the pilot does not fatigue in a prolong application. Table 5.3 shows the maximum allowable longitudinal control forces for FAR 23 and FAR 25 (as part of *handling qualities* requirements).

As a side note, in pilot training, pilot language, and some flight manuals, lift force on the aft tail (and force on the control column) is referred to as the back pressure. This is technically not correct, since pressure and force have different units.

Table 5.3 Maximum allowable longitudinal control forces

No.	Application	Maximum allowable longitudinal control force (lb)	
		FAR 23	FAR 25
1	For temporary force application		
	a. Center stick	60	No requirement
	b. Wheel	75	75
2	For prolong application	10	10

In aircraft that are equipped with fly-by-wire (FBW) control system (e.g., Lockheed F-22 Raptor, Eurofighter Typhoon, Airbus 320, and Airbus 380), a side-stick (instead of regular yoke/stick) may be employed. Sidesticks are shorter in height and provide less visual feedback to the pilots. The stick to elevator gearing ratio (G_e) for sidestick is coming from FBW control law, rather than Eq. 5.62.

Example 5.4 An aircraft with a weight of 5000 lb and a wing area of 260 ft^2 is cruising at sea level altitude with a speed of 150 knot. The aircraft has 10° of angle of attack and employs $-$ 20° of elevator deflection in order to maintain longitudinal trim. Other data of the aircraft are given below:

$$S_e = 1.4 \text{ ft}^2; C_e = 2 \text{ in}; i_h = -1 \text{ deg}; \varepsilon = 0; G_e = 0.8 \frac{\text{rad}}{\text{ft}};$$

$$C_{ho} = 0; C_{h_{\alpha_h}} = -0.21 \frac{1}{\text{rad}}; C_{h_{\delta E}} = -1.8 \frac{1}{\text{rad}}$$

Determine elevator hinge moment and stick force for this flight condition.

Solution

$$\alpha_h = \alpha + i_h - \varepsilon = 10 - 1 + 0 = 9 \text{ deg} \tag{5.58}$$

$$C_h = C_{ho} + C_{h_{\alpha_h}} \alpha_h + C_{h_{\delta E}} \delta_E = 0 - 0.21 \times \frac{9}{57.3} - 1.8 \times \frac{-20}{57.3} = 0.6 \tag{5.64}$$

$$H = \frac{1}{2} \rho U_1^2 S_E C_E C_h = \frac{1}{2} \times 0.002378 \times (150 \times 1.688)^2 \times 1.4 \times \frac{2}{12} \times 0.6 = 10.6 \text{ lb.ft} \tag{5.63}$$

$$F_s = G_e H = 0.8 \times 10.6 = 8.47 \text{ lb} \tag{5.68}$$

This level of required force is within the limit of a regular pilot force (power).

5.6.3 Trim Tab

In terms of stick force and hinge moment, majority of aircraft have two control modes: (1) Stick-fixed, and (2) Stick-free. In the *stick-fixed* mode, the stick should be kept by pilot's hand (Fig. 5.9), to keep the control surface (e.g., elevator) in place. In the *stick-free* mode, the stick is left by pilot to move freely. However, in order to keep the elevator in place, there are various means. In some aircraft, elevator tab is employed, so that, the stick can be left to stay free. When stick is free to move ($F_s = 0$), the elevator is free to move too. However, the elevator tab keeps the elevator in place, by trimming the pitching moment about elevator hinge (H = 0).

To allow the stick to be free, pilot first needs to select the deflection setting of the elevator by hand. Then, leave it to the tab to maintain the elevator deflection. The concept of stick-free and stick-fixed may be applied to rudder and aileron too. In order to control the aircraft in the stick-free mode (hands-off), the control surface should be aerodynamically balanced.

$$H = 0 \Rightarrow C_h = 0 \tag{5.71}$$

One method to aerodynamically balance a control surface is by employing a tab at the trailing edged of the control surface. Tabs are referred to as the secondary control surfaces and are placed at the trailing edges of the primary control surfaces. There are various tabs used in aircraft, examples are: (1) Trim tab, (2) Spring tab, (3) servo tab, 4. Balance tab, and 5. Anti-balance tab.

The most basic tab is a trim tab (see Fig. 5.10). As the name implies, it is used on elevators to longitudinally trim the aircraft in a cruising flight. Trim tabs are used to reduce the force the pilot applies to the stick to zero. Tab ensures that the pilot will not fatigue for holding the stick/yoke/wheel in a prolonged flight.

Trim tabs are frequently used in reversible flight control systems (e.g., mechanical). Trim tabs are utilized even in very large transport aircraft (e.g., Boeing 777); due to the fact that, the loss of all engines is conceivable; thus, pilot must be able to control and trim the jumbo aircraft with available his/her body force.

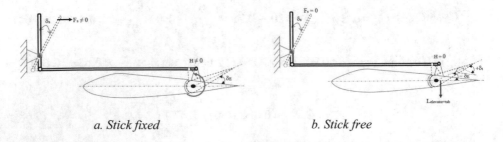

a. Stick fixed   b. Stick free

Fig. 5.9 Stick free and stick fixed

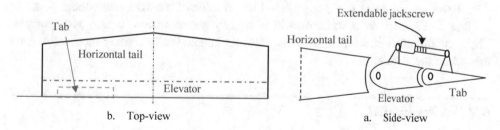

Fig. 5.10 Trim tab

For instance, in the past history flight of Boeing 747, there is at least three cases where all four engines become inoperative. These incidents demonstrate the necessity for providing an alternate manual control of control surfaces, even in large transport aircraft. Aircraft with powered flight controls need trim tabs to employ, during reversion to mechanical back up, if the hydraulic power transmission mechanism fails.

To achieve a zero cockpit control force, the trim tab is deflected opposite to the elevator deflection. When a lifting surface (e.g., tail) is equipped with a tab, the hinge moment coefficient; C_h is given by:

$$C_h = C_{h_o} + C_{h_\alpha}\alpha_{LS} + C_{h_{\delta_c}}\delta_c + C_{h_{\delta_t}}\delta_t \qquad (5.72)$$

where δ_t represents the tab deflection, and the parameter $C_{h_{\delta_t}}$ is a non-dimensional derivative as follows:

$$C_{h_{\delta_t}} = \frac{\partial C_h}{\partial \delta_t} \qquad (5.73)$$

The effectiveness of the tab ($C_{h_{\delta_t}}$) is a function of tab geometry and tab hinge line. There are two hinges in a lifting surface that its control surface possesses a tab; one for control surface deflection, and one for tab deflection. Trim tabs may be adjusted when the aircraft is on the ground; or may be manually operated and set by the pilot during flight. Trim tabs are usually deflected (by pilot at the cockpit) via a device referred to as the *trim wheel*. Trim wheel and trim tab allow a pilot to longitudinally trim a large aircraft with his/her hand force, and to keep a large elevator at any deflection.

The tab deflection is proportional to the control surface deflection. The tab-to-control-surface-chord ratio is usually about 0.2–0.4. Large transport aircraft such as Boeing 747 and Boeing 737 are equipped with elevator trim tab, as well as the rudder trim tab. The rudder tab for a Boeing 737 is about 3 m by 1.3 m. It is hinged to the rudder and operated mechanically by three rods. The tab deflects about twice as far as the rudder. The rudder moves ± 27 deg (left and right), while the tab moves ± 54 deg (right and left).

When the stick is released (i.e., hands-off), that is, the elevator is set free, the stability and control characteristics of the aircraft are impacted. For instance, the location of the stick-fixed neutral point is slightly behind the location of the stick-free neutral point. In another word, stick fixed static margin is slightly greater than stick-free static margin. Thus, an aircraft in stick-fixed mode is statically longitudinally slightly more stable than that when the stick is free.

5.7   Problems

1. Consider a transport aircraft is cruising with a speed of 195 knot and an angle of attack of $2°$ and the following longitudinal transfer functions:

$$\frac{u(s)}{\delta_E(s)} = \frac{-1000s^3 + 50{,}000s^2 + 500{,}000s + 10}{300s^4 + 3000s^3 + 20{,}000s^2 + 900s + 600}$$

$$\frac{\alpha(s)}{\delta_E(s)} = \frac{-60s^3 - 6000s^2 - 600s - 500}{300s^4 + 3000s^3 + 20{,}000s^2 + 900s + 600}$$

$$\frac{\theta(s)}{\delta_E(s)} = \frac{-7000s^2 - 15{,}000s - 1000}{300s^4 + 3000s^3 + 20{,}000s^2 + 900s + 600}$$

The pilot applies $-1°$ of elevator deflection. A. Plot the variations of angle of attack, airspeed, and pitch angle as the aircraft response to this input. B. Describe the aircraft new longitudinal trim point.

2. Consider a fixed-wing General Aviation aircraft with a maximum takeoff weight of 2200 lb, a wing planform area of 160 ft^2, is cruising with a speed of 95 knot at 10,000 ft. Other characteristics are:

$$C_{L_{\alpha_h}} = 5.1 \, \frac{1}{rad}; \, C = 4.5 \, ft; \, \eta_h = 0.96; \, l_h = 15 \, ft; \, S_h = 42 \, ft^2; \, I_{yy} = 1260 \, slug.ft^2$$

The elevator chord is 25% of the horizontal tail chord. Pilot applies a $-15°$ of elevator deflection. Determine the angular pitch acceleration ($\ddot{\theta}$) as a response of the aircraft to this input.

3. Consider a GA aircraft with a weight of 2700 lb and a wing area of 190 ft^2 and the following stability and control derivatives:

$$X_\alpha = 29 \, \frac{ft}{s^2}; \, X_u = -0.04 \frac{1}{s}; \, X_{\delta_E} = -0.7 \frac{ft}{s^2}$$

$$M_\alpha = -33 \, \frac{1}{s^2}; \, M_u = 0.0002 \frac{1}{ft.s}; \, M_q = -3.1 \, \frac{1}{s}; \, M_{\dot{\alpha}} = -1.6 \frac{1}{s}; \, M_{\delta_E} = -24 \frac{1}{s^2}$$

$$Z_\alpha = -395 \, \frac{ft}{s^2}; \, Z_u = -0.4 \frac{1}{s}; \, Z_q = -9.5 \, \frac{ft}{s}; \, Z_{\dot{\alpha}} = -1.3 \frac{ft}{s}; \, Z_{\delta_E} = -28 \frac{ft}{s^2}$$

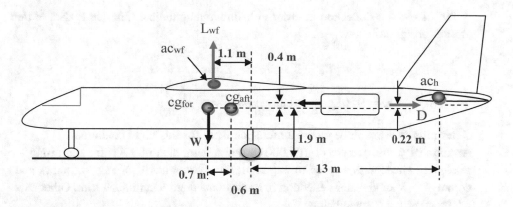

Fig. 5.11 Geometry of aircraft for Problem 4

The aircraft is cruising with an airspeed of 110 knot and an angle of attack of 3° at sea level. The pilot applies − 3° of elevator deflection. A. Plot the variations of angle of attack, airspeed, and pitch angle as the aircraft response to this input. B. Describe the aircraft new longitudinal trim point.

4. Figure 5.11 illustrates the geometry of a high-wing twin jet engine light utility aircraft which is equipped with a tricycle landing gear. The aircraft has the following characteristics.

$$m_{TO} = 25{,}000\,\text{kg},\ V_s = 91\,\text{knot},\ I_{yy} = 200{,}000\,\text{kg m}^2,\ T_{max} = 2 \times 35\,\text{kN},$$
$$V_c = 390\,\text{knot (at 25,000 ft)},\ C_{L_O} = 0.3,\ C_{D_{oC}} = 0.022,\ C_{D_{oTO}} = 0.041$$

Wing:

$$S = 93\,\text{m}^2,\ AR = 9,\ e = 0.9,\ \lambda = 1,\ \Delta C_{L_{flap}} = 0.8,$$
$$C_{m_{ac_{wf}}} = 0.04,\ i_w = 3\,\text{deg},\ h_o = 0.26$$

Horizontal tail:

$$S_h = 22\,\text{m}^2,\ b_h = 11\,\text{m},\ C_{L_{\alpha_h}} = 4.5\ 1/\text{rad},\ i_h = -1.5\,\text{deg},$$
$$\lambda_h = 1,\ \eta_h = 0.95,\ \text{Airfoil: NACA 0009},$$
$$\varepsilon = 1\,\text{deg},\ C_E/C_h = 0.26$$

How much elevator must be deflected at the take-off rotation to create a pitch angular acceleration of 10 deg/s²?

5. An aircraft with a weight of 6000 lb and a wing area of 290 ft² is cruising at sea level altitude with a speed of 120 knot. The aircraft has 6° of angle of attack and employs

− 20° of elevator deflection in order to maintain longitudinal trim. Other data of the aircraft are given below:

$$S_e = 1.8\,\text{ft}^2; \; C_e = 3 \text{ in}; \; i_h = -2\,\text{deg}; \; \varepsilon = 0; \; G_e = 0.7\,\frac{\text{rad}}{\text{ft}};$$

$$C_{ho} = 0; \; C_{h\alpha_h} = -0.25\frac{1}{\text{rad}}; \; C_{h\delta e} = -2.1\frac{1}{\text{rad}}$$

Determine elevator hinge moment and stick force for this flight condition.

6. An aircraft with a weight of 100,000 lb and a wing area of 1200 ft^2 is cruising at sea level altitude with a speed of 320 knot. The aircraft has 4° of angle of attack and employs − 5° of elevator deflection in order to maintain longitudinal trim. Other data of the aircraft are given below:

$$S_e = 360\,\text{ft}^2; \; C_e = 3.6 \text{ ft}; \; i_h = -1.4\,\text{deg}; \; \varepsilon = 1.1\,\text{deg};$$

$$G_e = 0.6\,\frac{\text{rad}}{\text{ft}}; \; C_{ho} = 0; \; C_{h\alpha} = -0.4\frac{1}{\text{rad}}; \; C_{h\delta e} = -2.6\frac{1}{\text{rad}}$$

Determine elevator hinge moment and stick force for this flight condition.

7. The following is the longitudinal state-space model of the fighter aircraft General Dynamics F-16 Fighting Falcon (Fig. 5.2) for the elevator input (δ_E), and two outputs (q and α):

$$\begin{bmatrix} \dot{q} \\ \dot{\alpha} \end{bmatrix} = \begin{bmatrix} -0.772 & -1.012 \\ 0.927 & -0.574 \end{bmatrix} \begin{bmatrix} q \\ \alpha \end{bmatrix} + \begin{bmatrix} -3.635 \\ -0.078 \end{bmatrix} \delta_E$$

Simulate and analyze the pitch response to an − 5° of elevator input. Plot the responses.

8. The following is the longitudinal state-space model of a GA for the elevator input (δ_E):

$$\begin{bmatrix} \dot{q} \\ \dot{\alpha} \\ \dot{u} \\ \dot{\theta} \end{bmatrix} = \begin{bmatrix} -1.4 & -104 & 0 & 0 \\ 0.99 & -2.7 & 0 & 0 \\ 0 & 0 & 0 & -32.2 \\ 1 & 0 & 0 & 0 \end{bmatrix} \begin{bmatrix} q \\ \alpha \\ u \\ \theta \end{bmatrix} + \begin{bmatrix} -15.8 \\ 0.22 \\ 0 \\ 0 \end{bmatrix} \delta_E \quad (5.15)$$

Simulate and analyze the pitch response to an − 3° of elevator input. Plot the responses.

9. Consider a fighter aircraft is cruising with a speed of 260 knot and an angle of attack of 1.2° and the following longitudinal transfer functions:

$$\frac{u(s)}{\delta_E(s)} = \frac{-400s^2 + 24,000s + 21,000}{300s^4 + 280s^3 + 600s^2 + 31s + 15}$$

$$\frac{\alpha(s)}{\delta_E(s)} = \frac{-25s^3 - 1400s^2 - 60s - 35}{300s^4 + 280s^3 + 600s^2 + 31s + 15}$$

$$\frac{\theta(s)}{\delta_E(s)} = \frac{-1500s^2 - 720s - 64}{300s^4 + 280s^3 + 600s^2 + 31s + 15}$$

The pilot applies $-2°$ of elevator deflection. A. Plot the variations of angle of attack, airspeed, and pitch angle as the aircraft response to this input. B. Describe the aircraft new longitudinal trim point.

10. Consider the following aircraft with a mass of 13,000 lb, a wing area of 230 ft^2 which is cruising at sea level with a speed of 230 knot:

$$C = 5 \text{ ft}; e = 0.9; AR = 10; \alpha = 2 \text{ deg};$$
$$C_{L_o} = 0.1; C_{D_o} = 0.024; C_{m_o} = -0.06;$$

$$I_{xx} = 28,000 \text{ slug.ft}^2; I_{yy} = 19,000 \text{ slug.ft}^2; I_{zz} = 48,000 \text{ slug.ft}^2;$$

$$C_{L_u} = 0.012 \frac{1}{\text{rad}}; C_{D_u} = 0.1 \frac{1}{\text{rad}}; C_{m_u} = 0.002 \frac{1}{\text{rad}};$$

$$C_{L_\alpha} = 5.2 \frac{1}{\text{rad}}; C_{D_\alpha} = 0.06 \frac{1}{\text{rad}}; C_{m_\alpha} = -1.8 \frac{1}{\text{rad}};$$

$$C_{L_{\dot\alpha}} = 1.6 \frac{1}{\text{rad}}; C_{D_{\dot\alpha}} = 0; C_{m_{\dot\alpha}} = -5.4 \frac{1}{\text{rad}};$$

$$C_{L_q} = 8.5 \frac{1}{\text{rad}}; C_{D_q} = 0; C_{m_q} = -15 \frac{1}{\text{rad}};$$

$$C_{L_{\delta_E}} = 0.4 \frac{1}{\text{rad}}; C_{D_{\delta_E}} = 0.012 \frac{1}{\text{rad}}; C_{m_{\delta_E}} = -1.3 \frac{1}{\text{rad}}$$

Determine aircraft longitudinal (elevator) transfer functions.

References

1. Stevens B. L., Lewis F. L., E. N. Johnson, Aircraft control and simulation, Third edition, John Wiley, 2016
2. Snell, A., F. Enns, and W. Garrard Jr, Nonlinear Inversion Flight Control for a Supermaneuverable Aircraft, *Journal of Guidance, Control and Dynamics*, Vol 15, No. 4, 1992.
3. Jackson P., Jane's All the World's Aircraft, Jane's information group, Various years
4. Sadraey M., Aircraft Design; A Systems Engineering Approach, Wiley, 2012

Lateral-Directional Control

6

6.1 Fundamentals

As discussed in Chap. 4, laterally/directionally motions are highly coupled, so the control of these two motions are discussed together. Control is defined as the process to changing the flight condition from one trim condition (e.g., cruise with an initial heading angle) to another trim condition (e.g., cruise with a new heading angle) with a desired rate or a specific duration. An aircraft must be laterally/directionally controllable at all flight conditions within the flight envelope. Missiles and fighter aircraft are laterally/directionally highly controllable, while aircraft such as hang gliders are lightly controlled.

The forces acting on an aircraft in flight consist of aerodynamic, thrust and gravitational forces. Any change in any of these forces will remove the trim condition and can be employed as a method to control the aircraft. In a fixed-wing aircraft, there are two lateral/directional aerodynamic control tools: (1) Aileron, and (2) Rudder.

All three aerodynamic forces of lift, drag, and side force, as well as two aerodynamic moments of rolling and yawing moments as contributing with various levels during any lateral-directional control process.

An important tool in lateral-directional stability and control is gyroscope which is a heavily weighted spinning disk that can maintain its orientation. It works on two principles: (1) Rigidity in space (i.e., if the gimbal rotates, the gyro will keep its direction), (2) Precession (i.e., tilting in response to a force). There are various types of gyroscopes. In modern aircraft, electric gyroscope is employed in turn coordination to show turn rate.

Lateral control is applied in the yz plane, while directional control is applied in the xy plane. Since these two control processes are highly coupled, we discussed them together in this chapter. The most challenging lateral-directional control is the turn, where all three control surfaces, all aerodynamic forces, and all aerodynamic moments are employed simultaneously. In this chapter, fundamentals, parameters, and governing equations for the aircraft lateral-directional control are presented. Other topics including coordinated turn, aileron reversal, and adverse yaw are explored too.

6.2   Lateral-Directional Governing Equations of Motion

Three contributing aerodynamic forces in lateral-directional control are: (1) Lift, (2) Drag, and (3) Side force. The other two contributing forces are: (4) Aircraft weight, and (5) Centrifugal force. There are two more contributors: (6) Aerodynamic rolling moment, and (7) Aerodynamic yawing moment. The aerodynamic control of an air vehicle about the roll, and yaw axes is accomplished by two control surfaces (Fig. 2.2) deflections: (1) Aileron (δ_A), and (2) Rudder (δ_R), respectively.

Five main flight parameters to control in a lateral-directional motion are: (1) Bank angle (ϕ), (2) Heading angle (ψ), (3) Roll rate (p), (4) Yaw rate (r), and (5) Sideslip angle (β). The following linear differential equations govern the variations of these parameters in a coupled lateral-directional motion:

$$m(\dot{v} + U_1 r) = Y + mg(\cos \Theta_1)\phi \tag{6.1}$$

$$I_{xx}\dot{p} - I_{xz}\dot{r} = L_A \tag{6.2}$$

$$I_{zz}\dot{r} - I_{xz}\dot{p} = N \tag{6.3}$$

$$\dot{\phi} = p \tag{6.4}$$

$$\dot{\psi} = r \tag{6.5}$$

The first three equations have been derived in Chap. 4. Equations (6.4) and (6.5) present the definitions of roll rate and yaw rate respectively.

As derived in Chap. 4, the lateral-directional aerodynamic forces and moment coefficients are modelled with respect to the dimensional stability and control derivatives. In terms of derivatives, Eqs. (6.1)–(6.3) are expanded and expressed as:

$$U_o\dot{\beta} + U_o\dot{\psi} = g\phi(\cos \Theta_o) + Y_\beta\beta + Y_p\dot{\phi} + Y_r\dot{\psi} + Y_{\delta_A}\delta_A + Y_{\delta_R}\delta_R \tag{6.6}$$

$$\ddot{\phi} - \frac{I_{xz}}{I_{xx}}\ddot{\psi} = L_\beta\beta + L_p\dot{\phi} + L_r\dot{\psi} + L_{\delta_A}\delta_A + L_{\delta_R}\delta_R \tag{6.7}$$

$$\ddot{\psi} - \frac{I_{Xz}}{I_{zz}}\ddot{\phi} = N_\beta\beta + N_p\dot{\phi} + N_r\dot{\psi} + N_{\delta_A}\delta_A + N_{\delta_R}\delta_R \tag{6.8}$$

Fig. 6.1 Lateral-directional
open-loop control system

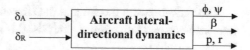

In a pure rolling motion, roll angular acceleration (rate of roll rate) is:

$$\ddot{\phi} = \dot{p} \tag{6.9}$$

In a pure yawing motion, yaw angular acceleration (rate of yaw rate) is:

$$\ddot{\psi} = \dot{r} \tag{6.10}$$

The applications of these equations are explored in the following sections. The open-loop control of an aircraft can be modeled by a block (see Fig. 6.1) with multiple inputs and multiple outputs (MIMO). The block contains the aircraft lateral-directional dynamic model with either: (1) Transfer function, (2) State space representation. If transfer functions are used, the impact of only one input (either aileron or rudder) will be investigated. However, if the state-space model is utilized, the impact of both inputs can be simultaneously analyzed.

Equations (6.6)–(6.10) can be converted (derivation is shown in Chap. 4) into state-space model as:

$$
\begin{bmatrix} \dot{\beta} \\ \dot{p} \\ \dot{r} \\ \dot{\phi} \end{bmatrix} =
\begin{bmatrix} \frac{Y_\beta}{u_o} & \frac{Y_p}{u_o} & -1 + \frac{Y_r}{u_o} & \frac{g\cos\theta}{u_o} \\ L_\beta & L_p & L_r & 0 \\ N_\beta & N_p & N_r & 0 \\ 0 & 1 & 0 & 0 \end{bmatrix}
\begin{bmatrix} \beta \\ p \\ r \\ \phi \end{bmatrix} +
\begin{bmatrix} \frac{Y_{\delta_A}}{U_o} & \frac{Y_{\delta_R}}{u_o} \\ L_{\delta_A} & L_{\delta_R} \\ N_{\delta_A} & N_{\delta_R} \\ 0 & 0 \end{bmatrix}
\begin{bmatrix} \delta_A \\ \delta_R \end{bmatrix} \tag{6.11}
$$

The output variables are tentatively selected to be the same as the state variables [β, p, r, ϕ]. Therefore, the output equation is presented as:

$$
\begin{bmatrix} \beta \\ p \\ r \\ \phi \end{bmatrix} =
\begin{bmatrix} 1 & 0 & 0 & 0 \\ 0 & 1 & 0 & 0 \\ 0 & 0 & 1 & 0 \\ 0 & 0 & 0 & 1 \end{bmatrix}
\begin{bmatrix} \beta \\ p \\ r \\ \phi \end{bmatrix} +
\begin{bmatrix} 0 & 0 \\ 0 & 0 \\ 0 & 0 \\ 0 & 0 \end{bmatrix}
\begin{bmatrix} \delta_A \\ \delta_R \end{bmatrix} \tag{6.12}
$$

In this lateral-directional state space model, A is a 4×4 square matrix, C is a 4×4 identity matrix, and B and D are 4×2 matrices, where in D, all elements are zero. Moreover, the transfer functions $\frac{\beta(s)}{\delta_A(s)}$, $\frac{\phi(s)}{\delta_A(s)}$, $\frac{\psi(s)}{\delta_A(s)}$, $\frac{\beta(s)}{\delta_R(s)}$, $\frac{\phi(s)}{\delta_R(s)}$, and $\frac{\psi(s)}{\delta_R(s)}$ have been already derived in Chap. 4 (Eqs. 4.39–4.44).

Either lateral-directional transfer functions or lateral-directional state space model can be employed to analyze state space model flight motions.

The ratio between the lift and the aircraft weight (W) is an important parameter in turn performance and turn control analysis. It is referred to as the load factor and is represented by the symbol n:

$$n = \frac{L}{W} \tag{6.13}$$

The load factor is a normalized representation of acceleration. In highly maneuverable fighter aircraft, the load factor may be as high as 10, while for high-speed supersonic missiles, it may pass 30.

6.3   Directional Control

6.3.1   Fundamentals

The directional control is the control of any linear/rotational motion in the x–y plane (e.g., yaw about z-axis, slipping, and skidding). Any change in the side-force and rolling and yawing moments (Eqs. 6.7 and 6.8) have the major influence on this control. Directional aerodynamic control of an aircraft can be achieved by providing an incremental lift force on the vertical tail. The yaw control is the heart of a directional control.

In a fixed-wing aircraft, the directional control (in the x–y plane) is performed through a directional control surface or rudder. In a fixed-wing conventional aircraft, the rudder (see Fig. 6.1) is part of the vertical tail (e.g., Boeing 777 and General Dynamics F-16 Fighting Falcon). The rudder—located at the trailing edge—is very similar to a flap, but it is deflected either left or right. By convention, a deflection of the rudder to the left is assumed as positive ($+ \delta_R$), which creates a positive side force (not shown in Fig. 6.1), and a negative aerodynamic yawing moment. A positive rudder deflection produces a positive vertical tail lift force ($+ L_v$), while a negative (counter-clockwise) yawing moment ($- \Delta N$) is generated. Thus, a positive side force (and a positive vertical tail lift) will produce a negative yawing moment.

In some fixed-wing aircraft (e.g., Northrop Grumman RQ-4 Global Hawk UAV), the rudder and elevator are combined, and referred to as the ruddervator. In these aircraft, the traditional horizontal and vertical tails are in the V-tail unconventional arrangement. The ruddervator will function as an elevator, when both pieces are deflected to opposite sides, while it will function as a rudder, when both pieces are deflected to the same sides. Thus, a positive rudder deflection turns the nose to the left.

When the rudder is deflected, a lift on vertical tail is created, which consequently generates a yawing moment (see Fig. 6.2). However, there is a strong coupling between lateral and directional state variables. Any lateral moment (L_A) will produce a directional motion (along y). Any directional moment (N) will produce a lateral motion (e.g., ϕ).

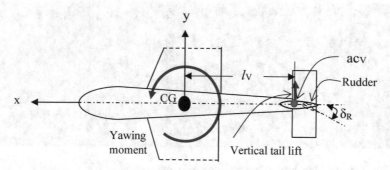

Fig. 6.2 Rudder deflection to create a yaw (top-view)

Hence, any aileron deflection, not only generates a rolling moment, but also creates a yawing moment. Moreover, any rudder deflection, not only generates a yawing moment, but also creates a rolling moment. In another word, any deflection in aileron (δ_A) or rudder (δ_R) will generates state variables of β, ϕ, ψ, P, R. The reason lies behind the existence of lateral moment arm and directional moment arm at the same time.

A change in the aircraft direction and heading is obtained via turn using rudder as well as the aileron, since a level turn is a combination of lateral and directional motions. The solution to adjust the relation between outputs of aileron and rudder is to follow a specific technique. To have a nice turn, the pilot needs to employ both rudder and aileron, such motion is referred to as the coordinated turn. The governing equations of a coordinated turn are presented in Sect. 6.5.

6.3.2 Directional Transfer Functions

The open-loop control of an aircraft can be modeled by a block (see Fig. 6.3) with multiple inputs and multiple outputs (MIMO). The block contains the aircraft directional dynamic model with either: (1) Transfer function, (2) State space representation. Three aircraft flight parameters which are directly controlled in a pure directional control are: (1) Heading angle (ψ), (2) Yaw rate (r), and (3) Sideslip angle (β). Other flight parameters which are also non-directly controlled in a directional motion include: (4) Bank angle (ϕ), and (5) Roll rate (p).

Fig. 6.3 Directional
open-loop control system

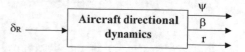

Fig. 6.4 The supersonic airliner Aérospatiale/BAC Concorde

In a fixed-wing conventional aircraft (e.g., Boeing 777 and Cessna Citation), the rudder is movable part of the vertical tail. The structure of a rudder—located at the tail trailing edge—is very similar to a wing flap, but it is deflected left and right.

In some large aircraft (e.g., Concorde, Fig. 6.4), rudder has two pieces, one upper and one lower (in tandem). Both pieces are employed at low speeds, while only the lower one is deflected at high speeds. This is to reduce the coupling between directional control with lateral control.

To analyze directional control, we need either: (1) Directional transfer functions, or (2) Lateral/Directional state space mathematical model. By applying LaPlace transform to the time-domain governing differential equations of motion (Eqs. 6.6–6.10), we can derive the directional transfer functions in s-domain.

The yaw-angle-to-rudder-deflection transfer function is:

$$\frac{\psi(s)}{\delta_R(s)} = \frac{A_{\psi R}s^3 + B_{\psi R}s^2 + C_{\psi R}s + D_2}{s\left[A_2s^4 + B_2s^3 + C_2s^2 + D_2s + E_2\right]} \tag{6.14}$$

The sideslip-angle-to-rudder-deflection transfer function is:

$$\frac{\beta(s)}{\delta_R(s)} = \frac{s\left[A_{\beta R}s^3 + B_{\beta R}s^2 + C_{\beta R}s + D_2\right]}{s\left[A_2s^4 + B_2s^3 + C_2s^2 + D_2s + E_2\right]} = \frac{A_{\beta R}s^3 + B_{\beta R}s^2 + C_{\beta R}s + D_2}{A_2s^4 + B_2s^3 + C_2s^2 + D_2s + E_2} \tag{6.15}$$

The bank-angle-to-rudder-deflection transfer function is:

$$\frac{\phi(s)}{\delta_R(s)} = \frac{s\left[A_{\phi R}s^2 + B_{\phi R}s + C_2\right]}{s\left[A_2s^4 + B_2s^3 + C_2s^2 + D_2s + E_2\right]} = \frac{A_{\phi R}s^2 + B_{\phi R}s + C_2}{A_2s^4 + B_2s^3 + C_2s^2 + D_2s + E_2} \tag{6.16}$$

Since yaw rate (r) is the determined by differentiation of yaw angle with time $(r(t) = \frac{d\psi(t)}{dt})$, the yaw-rate-to-rudder-deflection transfer function is obtained by multiplying numerator of $\frac{\psi(s)}{\delta_R(s)}$ by an "s" (i.e., in s-domain, R(s) = s ψ (s)). The yaw-rate-to-rudder-deflection transfer function is:

$$\frac{r(s)}{\delta_R(s)} = \frac{s\left(A_{\psi R}s^3 + B_{\psi R}s^2 + C_{\psi R}s + D_2\right)}{s\left[A_2s^4 + B_2s^3 + C_2s^2 + D_2s + E_2\right]} = \frac{A_{\psi R}s^3 + B_{\psi R}s^2 + C_{\psi R}s + D_2}{A_2s^4 + B_2s^3 + C_2s^2 + D_2s + E_2}$$

(6.17)

The coefficients $A_{\psi R}$ through E_2 have been introduced in Chap. 4. All directional transfer functions have the same fifth-order denominator (i.e., characteristic equation). However, for non-ψ transfer functions, the parameter "s" from both numerator and denominator can be removed.

The directional control gets harder, as: (1) Aircraft gets heavier (keeping geometry), (2) Aircraft cg moves forward, and (3) Aircraft mass moment of inertia about z-axis (I_{zz}) is increased.

Example 6.1 A single engine GA aircraft with a weight of 2600 lb and a wing area of 170 ft^2 is cruising at sea level with a speed of 80 knot. The aircraft has the following sideslip-angle-to- and yaw-angle-to-rudder-deflection transfer functions for this flight condition:

$$\frac{\beta(s)}{\delta_R(s)} = \frac{19s^3 + 2500s^2 + 29{,}700s - 510}{220s^4 + 3160s^3 + 6200s^2 + 30{,}260s + 540}$$

$$\frac{\psi(s)}{\delta_R(s)} = \frac{-\left(2200s^3 + 29{,}700s^2 + 2800s + 8500\right)}{s\left(220s^4 + 3160s^3 + 6200s^2 + 30{,}260s + 540\right)}$$

The pilot deflects rudder with the amount of $+ 5°$. Analyze the yaw response of the aircraft to this step input in rudder deflection.

Solution

To simulate the yaw and to plot the response history, the following matlab code is written:

```
b_dR=tf([19 2500 29700 -510],[220 3160 6200 30260 540])
psi_dR=tf([-2200 -29700 -2800 -8500],[220 3160 6200 30260 540 0])
t1=0:0.05:5;
dR=0*t1+5;
[beta,t]=lsim(b_dR,dR,t1);
[psi,t]=lsim(psi_dR,dR,t1);
subplot (2,1,1)
plot(t,beta,'ro'); grid
ylabel('\beta (deg)')
xlabel('Time (sec)')
title('The side slip angle response to +5° of rudder deflection')
subplot (2,1,2)
plot(t,psi,'b*'); grid
```

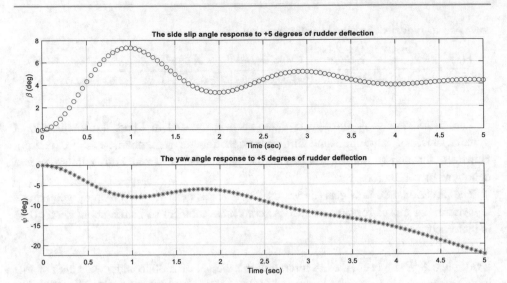

Fig. 6.5 Yaw response to + 5° of rudder deflection

```
ylabel('\psi (deg)')
xlabel('Time (sec)')
title('The yaw angle response to +5° of rudder deflection')
```

When this code is executed (simulation for 5 s), the plots shown in Fig. 6.5 are generated.

This plot indicates that the sideslip angle is oscillating but will have a steady state value of 4°. Moreover, the yaw angle is continuously increased, and after five seconds, it reaches about − 22°. Please note that, in this example, we only observe the aircraft pure directional motion. In reality, this motion will impact the lateral motion and will have bank angle response too. In Sect. 6.5, the simultaneous of coupled lateral-directional motions will be discussed.

6.3.3 Yaw Control

The aerodynamic yawing moment (N) is primarily a function of vertical tail lift (L_{vt}) and the tail arm (l_v). We shall examine how a rudder on a vertical tail provides the required control moment. This moment is equal to the vertical tail lift multiplied by tail arm (i.e., the distance between vertical tail ac to aircraft cg; l_v).

$$\Delta N = \Delta L_v l_v \tag{6.18}$$

This moment can also be calculated directly from rudder deflection:

$$\Delta N = \frac{1}{2}\rho V^2 Sb(\Delta C_n) \tag{6.19}$$

where the change in yawing moment coefficient is:

$$\Delta C_n = C_{n_{\delta_R}} \Delta \delta_R \tag{6.20}$$

The parameter $C_{n_{\delta_R}}$ is called the non-dimensional directional control power derivative or rudder control power derivative. The larger the value of $C_{n_{\delta_R}}$, the more effective the control is in creating the directional control moment. Note that, the parameter $\Delta \delta_R$ is employed, which implies, it is a deflection beyond what is needed for directional trim.

$$\delta_{R_{new}} = \delta_{Ro} + \Delta \delta_R \tag{6.21}$$

where δ_{Ro} is the initial elevator deflection that was used to directionally trim the aircraft, before a control desire is applied. The directional control moment produces an angular acceleration about z-axis, or yaw rate:

$$N = I_{zz}\ddot{\psi} \tag{6.22}$$

This equation is reformatted by using dimensional control derivatives:

$$N_{\delta_R}\delta_R = \ddot{\psi} = \dot{r} \tag{6.23}$$

The parameter N_{δ_R} is a control derivative and is called the dimensional rudder control power derivative, which is obtained by:

$$N_{\delta_R} = \frac{\overline{q_1}SbC_{n_{\delta_R}}}{I_{zz}} \tag{6.24}$$

where non-dimensional derivative $C_{n_{\delta_R}}$ is the rate of change of yawing moment coefficient due to rudder deflection $\left(\frac{\partial C_n}{\partial \delta_R}\right)$ and is obtained by:

$$C_{n_{\delta_R}} = -C_{L_{\alpha_V}} \overline{V}_V \eta_V \tau_r \tag{6.25}$$

where τ_r is the angle of attack effectiveness of the rudder (represents the rudder effectiveness) and is determined from Fig. 5.3 (in Chap. 5). The typical value for derivative $C_{n_{\delta_R}}$ is from 0 to $- 0.15$ 1/rad.

The rudder directional control power is a function of: (1) Rudder chord, (2) Rudder span, (3) Rudder deflection, (4) Vertical tail arm, (5) Vertical tail planform area, (6) Vertical tail airfoil, and (7) Vertical tail efficiency. As any of these parameters increased/improved, the rudder will be more effective is yaw control. The directional control gets harder, as: (1) Aircraft gets heavier (keeping geometry), (2) Aircraft cg moves forward, and (3) Aircraft mass moment of inertia about z-axis (I_{zz}) is increased. The derivative $C_{n_{\delta_R}}$ can be used for sizing the rudder to satisfy directional control requirement.

Example 6.2 Consider a fixed-wing General Aviation aircraft with a maximum takeoff weight of 2200 lb, a wing planform area of 160 ft^2, is cruising with a speed of 100 knot at 10,000 ft where air density is 17.56×10^{-4} slug/ft^3. Other characteristics are:

$$C_{L_{\alpha_v}} = 5.1 \frac{1}{rad}; b = 36 \text{ ft}; \eta_v = 0.96; l_v = 15 \text{ ft}; S_v = 35 \text{ ft}^2; I_{zz} = 2000 \text{ slug.ft}^2$$

The rudder chord is 30% of the vertical tail chord. The pilot applies a $-15°$ of rudder deflection. Determine the initial angular yaw acceleration ($\ddot{\psi}$) as a response of the aircraft to this input.

Solution

$$\overline{V}_V = \frac{l_v S_v}{bS} = \frac{15 \times 35}{36 \times 160} = 0.091 \tag{2.38}$$

From Fig. 5.3, for $C_R/C_V = 0.3$, we obtain $\tau_r = 0.52$.

$$C_{n_{\delta_R}} = -C_{L_{\alpha_V}} \overline{V}_V \eta_V \tau_r = -5.1 \times 0.091 \times 0.96 \times 0.52 = -0.232 \frac{1}{rad} \tag{6.25}$$

The change in yawing moment coefficient is:

$$\Delta C_n = C_{n_{\delta_R}} \Delta \delta_R = -0.232 \times \frac{-15}{57.3} = 0.061 \tag{6.20}$$

The yawing moment:

$$\Delta N = \frac{1}{2} \rho V^2 Sb(\Delta C_n) = \frac{1}{2} \times 0.001756 \times (100 \times 1.688)^2 \times 160 \times 36 \times (0.061) \tag{6.19}$$

$$\Delta N = 8752 \text{ lbf ft}$$

This is a positive (nose to the right) yawing moment, as we expected, due to negative (i.e., to the right) rudder deflection.

$$\Delta N = I_{zz}\dot{r} => \dot{r} = \ddot{\psi} = \frac{\Delta N}{I_{zz}} = \frac{8752}{2000} = 4.38 \frac{rad}{s^2} = 251 \frac{deg}{s^2} \tag{6.22}$$

6.3.4 Rudder Effectiveness

Directional control of an aircraft via rudder happens in a number of cases: (1) Asymmetric thrust, (2) Crosswind landing, (3) Spin recovery, (4) Coordinated turn, and (5) Adverse yaw. In asymmetric thrust, rudder will provide directional control/trim, when one engine of a multi-engine aircraft—is inoperative. In a crosswind landing, rudder will allow an aircraft to maintain alignment with the runway when a crosswind is blowing.

In a spinnable aircraft, the rudder has the task to help the aircraft to oppose the spin rotation and to recover from a spin. The functions of rudder in coordinating a turn and overcoming the adverse yaw—that is produced by ailerons—will be discussed in separate sections in this chapter.

When the rudder is deflected, it changes two aerodynamic forces and one aerodynamic moment: (1) Aircraft total side force, (2) Vertical tail lift, and (3) Aircraft yawing moment. The new values of these forces and moment will cause the aircraft to move a new directional trim point. A rudder deflection will create a constant side force but will cause the yawing moment to go from a maximum value to zero in a short time. The rate of transition from the initial longitudinal trim point to a new trim point (i.e., yaw rate) is a function of rudder effectiveness.

The aircraft new directional trim point—in a purely directional motion—features a new sideslip angle. Due to a non-zero sideslip angle (and consequently non-zero side force), the aircraft heading angle shall continuously vary. Note that, in a real flight, the bank angle and a number of longitudinal flight parameters will continuously vary too. If the change of bank angle is not controlled (i.e., limited), the aircraft will eventually enter a spin and dive.

To examine how a rudder provides the required directional control moment, we can evaluate three non-dimensional control derivatives: (1) $C_{n_{\delta_R}}$, (2) $C_{y_{\delta_R}}$, and (3) $C_{L_{v_{\delta_R}}}$. The significance of the derivative $C_{n_{\delta_R}}$—which represents the rate of change of yawing moment coefficient due to rudder deflection—is discussed earlier in this section. The higher the value of $C_{n_{\delta_R}}$, the more effective is the rudder. The representation of rudder effectiveness by other two derivatives are discussed in this section.

The rudder has the following main geometric characteristics: (1) rudder-chord-to-vertical-tail-chord ratio (C_R/C_v), (2) rudder-span-to-vertical-tail-span ratio (b_R/b_v), and (3) Maximum rudder deflection ($\pm\delta_{R_{max}}$). The rudder span is often selected to be equal to the vertical tail span (i.e., $b_R/b_v = 1$). The rudder chord ratio is often selected to be either about 20–30% (for GA and transport aircraft), or equal to 1 (for fighters). In fighter aircraft, an all moving vertical tail ($C_R = C_v$) is utilized to satisfy high directional maneuverability requirements. Table 6.1 shows specifications of rudders [1] for several aircraft.

Table 6.1 Specifications of rudders for several aircraft

No.	Aircraft	Type	m_{TO} (kg)	S_R/S_V	C_R/C_V	δ_{Rmax} (deg)
1	Cessna 182	Light GA	1406	0.38	0.42	±24
2	Cessna 650	Business jet	9979	0.26	0.27	±25
3	Gulfstream 200	Business jet	16,080	0.3	0.32	±20
4	Air Tractor AT-802	Regional airliner	18,600	0.61	0.62	±24
5	Lockheed C-130E Hercules	Military cargo	70,305	0.239	0.25	±35
6	DC-8	Transport	140,600	0.269	35	±32.5
7	DC-10	Transport	251,700	0.145	38	±23/±46[a]
8	Boeing 737-100	Transport	50,300	0.25	0.26	±27
9	Boeing 777-200	Transport	247,200	0.26	0.28	±27.3
10	Boeing 747-200	Transport	377,842	0.173	0.22	±25
11	Lockheed C-5A	Cargo	381,000	0.191	0.2	±25
12	Embraer ERJ145	Regional jet	22,000	0.29	0.31	±15
13	Airbus A340-600	Airliner	368,000	0.31	0.32	±31.6
14	General Dynamics F-16	Fighter	19,200	1	1	±30

[a]Tandem rudder

6.3.5 Minimum Control Speed

An important safety concern for multi-engine aircraft regarding directional control is the lowest speed that an aircraft can be directionally controlled, when One Engine is Inoperative (OEI). In the case of one (or more) engine(s) inoperative, an asymmetric yawing moment is generated by operating engines (Fig. 6.6):

$$N_{asym} = \sum T_i y_{T_i} \tag{6.28}$$

where T_i is the thrust of ith operative engine and y_{Ti} is the distance between ith engine thrust and the aircraft center of gravity in y direction. This moment can only be nullified or trimmed via an aerodynamic yawing moment (by deflecting rudder). In a directionally trimmed aircraft, the sum of the yawing moments about the aircraft center of gravity should be zero:

$$\sum N_{cg} = 0 \tag{6.29}$$

The aerodynamic yawing moment is the product of a vertical tail lift and its distance to aircraft center of gravity (as the moment arm, l_{vt}). The vertical tail lift, L_{vt} is:

Fig. 6.6 Yawing moment for directional control in an OEI situation (top view)

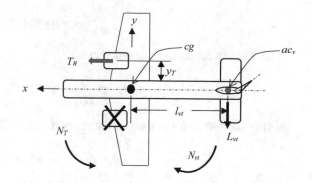

$$L_{vt} = \frac{1}{2}\rho V^2 S_{vt} C_{L_{vt}} \tag{6.30}$$

where S_{vt} is the vertical tail planform area (including rudder) and C_{Lvt} is the vertical tail lift coefficient. This equation demonstrates that the vertical tail lift is a function of aircraft speed, vertical tail area, air density, and vertical tail lift coefficient.

In an aircraft with a fixed configuration at a specific elevation, the only variable that can be varied is aircraft speed. Indeed, the maximum vertical lift coefficient is obtained by maximum rudder deflection which its typical value is about 1.2–1.6. When operating engines are generating their maximum thrusts, the asymmetric thrust moment is known. A directional control/trim is achieved when the vertical tail lift can produce the yawing moment (N_{vt}) equal to the thrust asymmetric yawing moment (N_{asym}).

$$N_{vt} = F_{vt} \cdot l_{vt} \tag{6.31}$$

where l_{vt} is the distance between vertical tail aerodynamic center to the aircraft center of gravity. Inserting Eqs. (6.28), (6.30), and (6.31) into directional trim equation (Eq. 6.29) allows us to obtain the minimum control speed as:

$$V_{mc} = \sqrt{\frac{2\sum T_i y_{T_i}}{\rho S_{vt} C_{L_{vt}} l_{vt}}} \tag{6.32}$$

If the aircraft speed is less than V_{mc}, while OEI, and if the pilot is not shutting down other engine(s), the aircraft will not be directionally controllable. In such condition, the aircraft will deviate out of runway, will hit an obstacle, and the aircraft will be damaged.

This V_{mc} should never be greater than stall speed, V_s; otherwise, the aircraft has a major design problem, and a significant safety issue. Such aircraft will not receive an FAA certificate and must be redesigned.

Another approach to determine the minimum control speed is to employ the aircraft aerodynamic yawing moment:

$$N = \frac{1}{2}\rho V^2 S C_n b \tag{1.18}$$

where the yawing moment coefficient is:

$$C_n = C_{n_\beta}\beta + C_{n_p} P \frac{b}{2U_1} + C_{n_r} R \frac{b}{2U_1} + C_{n_{\delta_a}}\delta_a + C_{n_{\delta_r}}\delta_r \tag{1.94}$$

For this condition, the yawing moment coefficient is reduced to:

$$C_n = C_{n_{\delta_r}}\delta_{r_{max}} \tag{6.33}$$

Thus, the minimum control speed is obtained as:

$$V_{mc} = \sqrt{\frac{2\sum T_i y_{T_i}}{\rho S b C_{n_{\delta_r}}\delta_{r_{max}}}} \tag{6.34}$$

Indeed, the drag of the inoperative engine will create an asymmetric yawing moment too. To include this moment, the numerator of Eqs. (6.32) and (6.34) is multiplied by a factor (k_{ND}).

$$V_{mc} = \sqrt{\frac{2k_{ND}\sum T_i y_{T_i}}{\rho S b C_{n_{\delta_r}}\delta_{r_{max}}}} \tag{6.35}$$

This factor (k_{ND}) is often about 1.1 (for aircraft with prop-driven engine with variable pitch; and aircraft with jet engine with low bypass ratio) to 1.2 (for aircraft with prop-driven engine with fixed pitch; and aircraft with jet engine with high bypass ratio). The aviation regulations require that the minimum control speed be related to the stall speed of the aircraft. FAA [11, 12] requires any GA and transport aircraft to satisfy the following requirement:

$$V_{mc} \leq 1.2 V_s \tag{6.36}$$

For military aircraft, MIL-F-8785C [2] requires that the aircraft to satisfy the following requirement:

$$V_{mc} \leq highest \ of \ 1.1 \ V_s \ or \ V_s + 10 \ knot \tag{6.37}$$

Example 6.3 A transport aircraft has twin high bypass ratio turbofan engines each generating 150 kN of thrust. The distance between each engine thrust line and aircraft center of gravity (along y-axis) is 10 m, and the distance between vertical tail aerodynamic center to the aircraft center of gravity (l_{vt}) is 40 m. If the maximum lift coefficient of the vertical tail is 1.4 and vertical tail area is 36 m^2, determine minimum control speed of this aircraft at sea level.

Solution

When one engine (say left one) is not operating, the right engine thrust (T_R) will produce an undesirable yawing moment that needs to be nullified. The minimum control speed can be determined as:

$$V_{mc} = \sqrt{\frac{2 k_{ND} \sum T_i y_{T_i}}{\rho S_{vt} C_{L_{vt}} l_{vt}}} = \sqrt{\frac{2 \times 1.2 \times (150{,}000 \times 10)}{1.225 \times 36 \times 1.4 \times 40}} = 38.2 \, \frac{\text{m}}{\text{s}} = 74.2 \text{ knot} \quad (6.35)$$

where k_{ND} is 1.2, since the aircraft is equipped with high bypass ratio jet engines.

6.3.6 Pure Yawing Motion Approximation

The approximate yawing moment equation can be derived from the approximation of the Dutch roll motion. We ignore any change in sideslip angle but consider only a change in yaw angle.

$$\ddot{\psi} - N_r \dot{\psi} + N_\beta \psi = N_{\delta_R} \delta_R \quad (6.38)$$

Applying the Laplace transform, this differential equation is readily converted to a transfer function:

$$s^2 \Psi(s) - N_r s \Psi(s) + N_\beta \Psi(s) = N_{\delta_R} . \delta_R(s) \quad (6.39)$$

or

$$\Psi(s) \left[s^2 - s N_r + N_\beta \right] = N_{\delta_R} . \delta_R(s) \quad (6.40)$$

Thus, the transfer function is obtained as:

$$\frac{\Psi(\text{s})}{\delta_R(\text{s})} = \frac{N_{\delta_R}}{s^2 - s N_r + N_\beta} \quad (6.41)$$

This is an approximate for the yaw-angle-to-rudder-deflection transfer function.

In Sect. 6.3, we mainly explored a pure directional control, and employed a linear directional dynamic model. However, in reality, directional motion is nonlinear and highly coupled with lateral motion. In Sect. 6.6, the fundamentals and control process and of a turn is discussed that includes both directional and lateral motions. For accurate analysis of directional and lateral motions, a nonlinear dynamic model and computer simulation are required.

6.4   Lateral Control

6.4.1   Governing Equations

The control of any rotational motion about the x-axis is called lateral control. In a fixed-wing aircraft (e.g., Airbus A-340) with a conventional configuration, the lateral motion is mainly executed using aileron deflection (spoiler may be used too). The roll control (i.e., change in the bank angle) is assumed as a lateral control. The aileron (see Fig. 2.2) is part of the wing, located at the outboard and trailing edge. An aileron—which are located on the outboard of the wing—is very similar to a flap, but it is deflected either up or down. It has two pieces, one on each side of the wing, and they are deflected differentially.

When the ailerons are deflected, the wing lift distribution is differentially changed. Any change in the wing lift distribution and rolling moment will have a major influence on this motion. The lift—on the section (say left, see Fig. 6.7) that aileron has been deflected downward—is increased; while the lift—on the section (say right) that aileron has been deflected upward—is decreased. This change in the wing lift distribution will consequently generates a rolling moment. Based on the right-hand rule, this is a positive roll, since the x-axis is coming out of the page.

In a pure rolling motion, a lateral control moment produces an angular acceleration about x-axis, or roll rate:

$$L_A = I_{xx}\ddot{\phi} = I_{xx}\dot{p} \tag{6.42}$$

The aircraft aerodynamic rolling moment is:

$$L_A = \frac{1}{2}\rho V^2 S C_l b \tag{1.19}$$

where the rolling moment coefficient in terms of non-dimensional derivatives is:

$$C_l = C_{l_\beta}\beta + C_{l_p}P\frac{b}{2U_1} + C_{l_r}R\frac{b}{2U_1} + C_{l_{\delta_a}}\delta_a + C_{l_{\delta_r}}\delta_r \tag{1.92}$$

The aerodynamic rolling moment in terms of dimensional derivatives is:

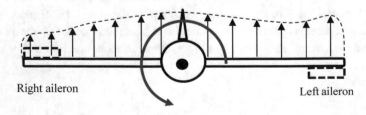

Right aileron   Left aileron

Fig. 6.7 Aileron deflection to create a roll (front-view)

$$\frac{1}{I_{xx}}\Delta L = L_\beta \beta + L_{\dot\beta}\dot\beta + L_p p + L_r r + L_{\delta_A}\delta_A + L_{\delta_R}\delta_R \tag{1.82}$$

In a pure rolling motion, the rolling moment and its coefficient are reduced to:

$$\frac{1}{I_{xx}}\Delta L = L_p p + L_{\delta_A}\delta_A \tag{6.43}$$

$$C_l = C_{l_p} P\frac{b}{2U_1} + C_{l_{\delta_a}}\delta_a \tag{6.44}$$

Inserting Eq. (6.43) into Eq. (6.42), the pure rolling motion governing equation will be:

$$L_p p + L_{\delta_A}\delta_A = I_{xx}\ddot\phi \tag{6.45}$$

The parameter L_{δ_A} is called the dimensional aileron control power derivative, and is obtained by:

$$L_{\delta_A} = \frac{\overline{q_1}SbC_{l_{\delta_A}}}{I_{xx}} \tag{6.46}$$

where non-dimensional derivative $C_{l_{\delta_A}}$ is the rate of change of rolling moment coefficient due to aileron deflection $\left(\frac{\partial C_l}{\partial \delta_A}\right)$. Reference [3] provides technique to determine this lateral control derivative.

The rate of change of rolling moment with aileron deflection is a sign of the aileron control effectiveness. As presented in Chap. 2, Ref. [4] provides an estimation technique to determine $C_{l_{\delta_A}}$ as:

$$C_{l_{\delta_A}} = \frac{2C_{L_{\alpha_w}}\tau_a C_r}{Sb}\left[\frac{y^2}{2} + \left(\frac{\lambda - 1}{b/2}\right)\frac{y^3}{3}\right]_{y_1}^{y_2} \tag{6.47}$$

where C_r is the wing root chord, λ is wing taper ratio (Eq. 2.48), τ_a is the aileron effectiveness parameter, and y_1 and y_2 are distances between inboard and outboard of aileron to the fuselage centerline. The derivative $C_{l_{\delta_A}}$ can be used for sizing the aileron to satisfy roll control requirement. The dimensional stability derivative L_p is introduced in Chap. 4 and reproduced here for convenience.

$$L_p = \frac{\overline{q_1}Sb^2 C_{l_p}}{2I_{xx}U_1} \tag{6.48}$$

Using definition of roll re ($\ddot\phi = \dot p$), we can rewrite Eq. (6.45) as:

$$\frac{-1}{L_p}\dot p(t) + p(t) = -\frac{L_{\delta_A}}{L_p}\delta_A(t) \tag{6.49}$$

This is a time-domain first order differential equation that governs the variations of roll rate (p) with time. The parameter $\frac{-1}{L_p}$ is referred to as the time constant (τ) of the rolling motion. The time constant is defined as the time required for the output parameter of a first-order stable dynamic system to reach to 63% of its final (steady-state) value, when a unit step input is applied.

The roll time constant is representing how fast the aircraft approaches a new steady-state roll rate after aileron deflected. If the roll time constant is small (e.g., a fraction of a second), the aircraft will respond very rapidly to an aileron input; while if the roll time constant is large (e.g., a few seconds), the aircraft will roll very slowly.

The solution of the differential Eq. (6.49) is:

$$p = -\frac{L_{\delta_A}}{L_p} \delta_A \left(1 - e^{L_p t}\right) \tag{6.50}$$

Recall that stability derivative L_P is negative. The steady state value of roll rate (p_{ss}) is obtained by using a time t that is large enough (t $\rightarrow \infty$) such that $e^{L_p t}$ is zero.

$$P_{ss} = -\frac{L_{\delta_A}}{L_P} \delta_A \tag{6.51}$$

Employing non-dimensional derivatives, Eq. (6.47) can be rewritten as:

$$P_{ss} = -\frac{C_{l_{\delta_A}}}{C_{l_p}} \frac{2u_o}{b} \delta_A \tag{6.52}$$

where u_o is the initial airspeed. When LaPlace transform is applied to Eq. (6.49), we obtain the following s-domain equation:

$$\frac{-1}{L_p} s P(s) + P(s) = -\frac{L_{\delta_A}}{L_P} \delta_A(s) \tag{6.53}$$

Then, the roll-rate-to-aileron-deflection transfer function is obtained as:

$$\frac{P(s)}{\delta_A(s)} = \frac{-\frac{L_{\delta_A}}{L_P}}{\frac{-1}{L_P} s + 1} \tag{6.54}$$

which can be further simplified to:

$$\frac{P(s)}{\delta_A(s)} = \frac{L_{\delta_A}}{s - L_P} \tag{6.55}$$

Using the definition of roll rate ($P = \dot{\phi}$), this equation can be employed to derive the bank angle transfer function, by integrating in s-domain (i.e., multiply by $\frac{1}{s}$).

$$\frac{\phi(s)}{\delta_A(s)} = \frac{1}{s} \frac{L_{\delta_A}}{s - L_P} = \frac{L_{\delta_A}}{s^2 - s L_P} \tag{6.56}$$

By applying inverse LaPlace transform, one can derive an expression to govern the variations of bank angle as a function of time.

$$\phi(t) = -\left(\frac{L_{\delta_A}\delta_A}{L_P}\right)t + \left(\frac{L_{\delta_A}\delta_A}{L_P^2}\right)\left(e^{L_P t} - 1\right) \quad (6.57)$$

The transfer functions (Eqs. 6.55 and 6.56) can be used for the simulation of a rolling motion. We can also determine the steady state value of roll rate (P_{ss}), by applying final value theorem (s $\rightarrow$ 0, when t $\rightarrow$ ∞) to any transfer function. For roll rate, we can write:

$$P_{t \rightarrow \infty} = \delta_A \left(\frac{L_{\delta_A}}{s - L_P}\right)_{s \rightarrow 0} = \frac{L_{\delta_A}}{0 - L_P}\delta_A \quad (6.58)$$

or

$$P_{ss} = -\frac{L_{\delta_A}}{L_P}\delta_A \quad (6.59)$$

Due to the selected sign convention, the lateral control derivative L_{δ_A} is always positive (i.e., $L_{\delta_A} > 0$), while lateral stability derivative L_P is always negative (i.e., $L_P < 0$).

Roll control requirements in terms of a bank angle change ($\Delta\phi$) in a given time (t) are specified by Ref. [5] for various aircraft classes. For instance, Table 6.2 provides roll *handling qualities* in terms of the time to achieve a specified bank angle change for Class I. For other classes, refer to [6].

Example 6.4 Consider a fighter aircraft with a weight of 30,000 lb and a wing area of 200 ft^2 is flying at sea level with a speed of 150 knot. The aircraft has a wing span of 22 ft, a mass moment of inertia (about x-axis) of 3500 slug.ft^2 and the following non-dimensional lateral control and control derivatives:

$$C_{l_{\delta_A}} = 0.06\frac{1}{\text{rad}}; C_{l_P} = -0.7\frac{1}{\text{rad}}$$

Calculate the roll response of the aircraft to a + 25° step input in aileron deflection.

Table 6.2 Roll control requirements for Class I

Level	Flight phase category		
	A	B	C
	Time to achieve a bank angle of 60° (s)	Time to achieve a bank angle of 45° (s)	Time to achieve a bank angle of 30° (s)
1	1.3	1.7	1.3
2	1.7	2.5	1.8
3	2.6	3.4	2.6

Solution

We first, calculate two lateral dimensional control and control derivatives:

$$\bar{q}_1 = \frac{1}{2}\rho U_o^2 = \frac{1}{2} \times 0.002378 \times (150 \times 1.688)^2 = 76.2 \frac{\text{lb}}{\text{ft}^2} \tag{1.23}$$

$$L_{\delta_A} = \frac{\bar{q}_1 Sb C_{l_{\delta_A}}}{I_{xx}} = \frac{76.2 \times 200 \times 22 \times 0.06}{3500} = 5.75 \frac{\text{rad}}{\text{s}^2} \tag{6.46}$$

$$L_P = \frac{\bar{q}_1 Sb^2 C_{l_p}}{2 I_{xx} U_1} = \frac{76.2 \times 200 \times 22^2 \times (-0.7)}{2 \times 3500 \times (150 \times 1.688)} = -2.914 \frac{\text{rad}}{\text{s}} \tag{6.48}$$

The steady state value of roll rate (P_{ss}) is:

$$P_{ss} = -\frac{L_{\delta_A}}{L_P}\delta_A = -\frac{5.75}{-2.914}\left(\frac{25}{57.3}\right) = 0.861 \frac{\text{rad}}{\text{s}} = 49.3 \frac{\text{deg}}{\text{s}} \tag{6.59}$$

Transfer functions:

$$\frac{P(s)}{\delta_A(s)} = \frac{L_{\delta_A}}{s - L_P} = \frac{5.75}{s - (-2.914)} \tag{6.55}$$

$$\frac{\phi(s)}{\delta_A(s)} = \frac{L_{\delta_A}}{s^2 - sL_P} = \frac{5.75}{s^2 - (-2.914s)} \tag{6.56}$$

To simulate the roll and to plot the roll time history, the following matlab code is written:

```
LP = -2.914;
LdA = 5.75;
P_dA=tf(LdA,[1 -LP]);
Phi_dA=tf(LdA,[1 -LP 0])
t1=0:0.05:2;
dA=0*t1+25;
[P1,t]=lsim(P_dA,dA,t1);
[Phi,t]=lsim(Phi_dA,dA,t1);
subplot (2,1,1)
plot(t,P1,'ro'); grid
ylabel('P (deg/sec)')
xlabel('Time (sec)')
title('The roll rate response to +25° of Aileron deflection')
subplot (2,1,2)
plot(t,Phi,'b*'); grid
ylabel('\phi (deg)')
xlabel('Time (sec)')
title('The bank angle response to +25° of Aileron deflection')
```

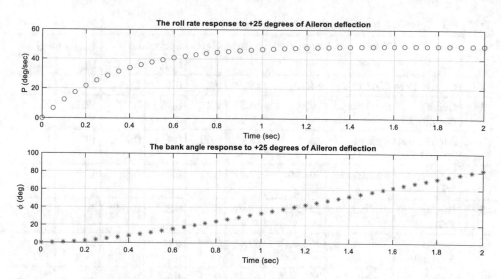

Fig. 6.8 The roll response to $+ 25°$ of Aileron deflection

When this code is executed (simulation for 2 s), the plots shown in (Fig. 6.8) are generated.

This plot confirms the steady state value of roll rate to be about 49 deg/s. Moreover, the bank angle is continuously increased, and after two seconds, the bank angle reaches about 81°. Please note that, in this example, we only observe the aircraft pure lateral motion. In reality, this motion will impact the directional motion and will have yaw angle response too. In Sect. 6.5, the simultaneous of coupled lateral-directional motions will be discussed.

6.4.2 Aileron Control Power

The aileron has the following main geometric characteristics: (1) aileron-chord-to-wing-chord ratio (C_A/C), (2) aileron-span-to-wing-span ratio (b_A/b_W), (3) aileron inboard (b_i) and outboard (b_o) span location (often about 60–95%), and (4) Maximum aileron deflection ($\pm\delta_{A_{max}}$). The aileron span often covers about 20% of the outer wing span (i.e., $b_A/b = 0.2$). The aileron chord ratio is often selected to be about 20–30%. Table 6.3 illustrates specifications of ailerons [1] for several aircraft.

Since aileron are used in pairs (has two pieces), and pieces may have different deflections, the aileron deflection is determined by finding the average:

$$\delta_A = \frac{1}{2}\left(\left|\delta_{A_R}\right| + \left|\delta_{A_L}\right|\right) \tag{6.60}$$

Table 6.3 Specifications of ailerons for several aircraft

No.	Aircraft	Type	m_{TO} (kg)	b (m)	C_A/C	Span ratio		δ_{Amax} (deg)	
						$b_i/b/2$	$b_o/b/2$	Up	Down
1	Cessna 182	Light GA	1406	11	0.2	0.46	0.95	20	14
2	Cessna Citation III	Business jet	9979	16.31	0.3	0.56	0.89	12.5	12.5
3	Air Tractor AT-802	Agriculture	7257	18	0.36	0.4	0.95	17	13
4	Gulfstream 200	Business jet	16,080	17.7	0.22	0.6	0.86	15	15
5	Boeing 777-200	Airliner	247,200	60.9	0.22	0.32[a]	0.76[b]	30	10
6	Airbus 340-600	Airliner	368,000	63.45	0.3	0.64	0.92	25	20
7	Airbus A340-600	Airliner	368,000	63.45	0.25	0.67	0.92	25	25
8	General Dynamics F-16	Fighter	19,200	9.96	0.21[c]	0.25	0.75	21	23

[a] Inboard aileron
[b] Outboard aileron
[c] Flaperon

A positive aileron deflection is one which results in a positive rolling moment. The aileron deflections range from 0 to 25°, while the maximum spoiler deflections range anywhere from 30 to 60°. When ailerons are deflected more than about 25°, a flow separation tends to occur. Beyond this value, aileron will lose its effectiveness. Furthermore, close to wing stall, even a small downward aileron deflection can produce separation, and loss of roll control effectiveness. To improve aileron control power, large transport aircraft employ spoiler.

Large transport aircraft (e.g., Boeing 787) also have a second pair of ailerons, so one "outboard aileron" and one "inboard aileron". Both pieces are employed at low speeds, while only the inner ones are deflected at high speeds. This is to improve aileron effectiveness and lateral control.

In some fighters, the flaps and ailerons are combined, such devices are referred to as *flaperons*. In most delta wing aircraft with (e.g., *Lockheed F-117 Nighthawk and Concorde*), the ailerons are combined with the elevators to form an *elevon* for pitch and roll control. Since, these aircraft have no horizontal tail, the elevons are placed along the trailing edge of the wing. Concorde (Fig. 6.4) is controlled in pitch and roll by 6 elevons (inner, middle, and outer pairs), and in yaw by 2 rudders.

The Aileron lateral control power is primarily represented by the dimensional aileron control power derivative L_{δ_A} or the non-dimensional aileron control power derivative $C_{l_{\delta_A}}$. As the values of these derivatives increased, the aileron is assumed to be stronger in lateral/roll control.

6.4.3 Lateral Transfer Functions

The open-loop lateral control of an aircraft can be modeled by a block (see Fig. 6.9) with multiple inputs and multiple outputs (MIMO). The block contains the aircraft lateral dynamic model with either: (1) Transfer function, (2) State space representation. Three aircraft flight parameters which are directly controlled in a pure lateral control are: (1) Bank angle (ϕ), (2) Roll rate (p) and (3) Sideslip angle (β). Other flight parameters which are also non-directly controlled in a lateral control process include: 4. Heading angle (ψ), and (5) Yaw rate (r).

To analyze lateral control, we need either: (1) Lateral transfer functions, or (2). Lateral/Directional state space mathematical model. By applying LaPlace transform to the time-domain governing differential equations of motion (Eqs. 6.6–6.10), we can derive the lateral transfer functions in s-domain.

The bank-angle-to-aileron-deflection transfer function is:

$$\frac{\phi(s)}{\delta_A(s)} = \frac{s\left[A_{\phi A}s^2 + B_{\phi A}s + C_{\phi A}\right]}{s\left[A_2s^4 + B_2s^3 + C_2s^2 + D_2s + E_2\right]} = \frac{A_{\phi A}s^2 + B_{\phi A}s + C_{\phi A}}{A_2s^4 + B_2s^3 + C_2s^2 + D_2s + E_2}$$

(6.61)

The sideslip-angle-to-aileron-deflection transfer function is:

$$\frac{\beta(s)}{\delta_A(s)} = \frac{s\left[A_{\beta A}s^3 + B_{\beta A}s^2 + C_{\beta A}s + D_{\beta A}\right]}{s\left[A_2s^4 + B_2s^3 + C_2s^2 + D_2s + E_2\right]} = \frac{+B_{\beta A}s^2 + C_{\beta A}s + D_{\beta A}}{A_2s^4 + B_2s^3 + C_2s^2 + D_2s + E_2}$$

(6.62)

The yaw-angle-to-aileron-deflection transfer function is:

$$\frac{\psi(s)}{\delta_A(s)} = \frac{A_{\psi A}s^3 + B_{\psi A}s^2 + C_{\psi A}s + D_{\psi A}}{s\left[A_2s^4 + B_2s^3 + C_2s^2 + D_2s + E_2\right]}$$

(6.63)

Fig. 6.9 Lateral open-loop control system

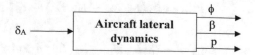

Since roll rate (p) is the determined by differentiation of bank angle with time $(p(t) = \frac{d\phi(t)}{dt})$, the roll-rate-to-aileron-deflection transfer function is obtained by multiplying numerator of $\frac{\phi(s)}{\delta_A(s)}$ by an "s" (i.e., in s-domain, $P(s) = s\,\phi\,(s)$). Thus, the roll-rate-to-aileron-deflection transfer function is:

$$\frac{P(s)}{\delta_A(s)} = \frac{s\left[A_{\phi A}s^2 + B_{\phi A}s + C_{\phi A}\right]}{A_2 s^4 + B_2 s^3 + C_2 s^2 + D_2 s + E_2} = \frac{A_{\phi A}s^3 + B_{\phi A}s^2 + C_{\phi A}s}{A_2 s^4 + B_2 s^3 + C_2 s^2 + D_2 s + E_2}$$

(6.64)

The coefficients A_2 through E_2 have been introduced in Chap. 4. All lateral transfer functions have the same fifth-order denominator (i.e., characteristic equation). However, for non-ψ transfer functions, the parameter "s" from both numerator and denominator can be removed.

The coefficients for bank-angle-to-aileron-deflection transfer function are:

$$A_{\phi A} = U_o\left(L_{\delta_A} + N_{\delta_A}\frac{I_{xz}}{I_{xx}}\right)$$

$$B_{\phi A} = U_o\left(N_{\delta_A}L_r - L_{\delta_A}N_r\right) - Y_\beta\left(L_{\delta_A} + N_{\delta_A}\frac{I_{xz}}{I_{xx}}\right) + Y_{\delta_A}\left(L_\beta + N_\beta\frac{I_{xz}}{I_{xx}}\right)$$

$$C_{\phi A} = -Y_\beta\left(N_{\delta_A}L_r - L_{\delta_A}N_r\right) + Y_{\delta_A}\left(L_r N_\beta - N_r L_\beta\right) + (U_o - Y_r)\left(N_\beta L_{\delta_A} - L_\beta N_{\delta_A}\right)$$

The coefficients for sideslip-angle-to-aileron-deflection transfer function are:

$$A_{\beta A} = Y_{\delta_A}\left(1 - \frac{I_{xz}}{I_{xx}}\frac{I_{xz}}{I_{zz}}\right)$$

$$B_{\beta A} = -Y_{\delta_A}\left(N_r + L_p + N_p\frac{I_{xz}}{I_{xx}} + L_r\frac{I_{xz}}{I_{zz}}\right) + Y_p\left(L_{\delta_A} + N_{\delta_A}\frac{I_{xz}}{I_{xx}}\right) + (Y_r - U_o)\left(L_{\delta_A}\frac{I_{xz}}{I_{zz}} + N_{\delta_A}\right)$$

$$C_{\beta A} = Y_{\delta_A}\left(N_r L_p - N_p L_r\right) + Y_p\left(N_{\delta_A}L_r - L_{\delta_A}N_r\right) + g\cos\theta_o\left(L_{\delta_A} + N_{\delta_A}\frac{I_{xz}}{I_{xx}}\right) + (Y_r - U_o)\left(L_{\delta_A}N_p - N_{\delta_A}L_p\right)$$

$$D_{\beta A} = g\cos\theta_o\left(N_{\delta_A}L_r - L_{\delta_A}N_r\right)$$

The coefficients for yaw-angle-to-aileron-deflection transfer function are:

$$A_{\psi A} = U_o\left(N_{\delta_A} + L_{\delta_A}\frac{I_{xz}}{I_{ZZ}}\right)$$

$$B_{\psi A} = U_o\left(L_{\delta_A}N_p - N_{\delta_A}L_p\right) - Y_\beta\left(N_{\delta_A} + L_{\delta_A}\frac{I_{xz}}{I_{zz}}\right) + Y_{\delta_A}\left(L_\beta\frac{I_{xz}}{I_{zz}} + N_\beta\right)$$

$$C_{\psi A} = -Y_\beta\left(L_{\delta_A}N_p - N_{\delta_A}L_p\right) + Y_p\left(N_\beta L_{\delta_A} - L_\beta N_{\delta_A}\right) + Y_{\delta_A}\left(L_\beta N_p - N_\beta L_p\right)$$

$$D_{\psi A} = g\cos\theta_o\left(N_\beta L_{\delta_A} - L_\beta N_{\delta_A}\right)$$

The coefficients for bank-angle-to-rudder-deflection transfer function are:

$$A_{\phi R} = U_o\left(L_{\delta_R} + N_{\delta_R}\frac{I_{xz}}{I_{xx}}\right)$$

$$B_{\phi R} = U_o\left(N_{\delta_R}L_r - L_{\delta_R}N_r\right) - Y_\beta\left(L_{\delta_R} + N_{\delta_R}\frac{I_{xz}}{I_{xx}}\right) + Y_{\delta_R}\left(L_\beta + N_\beta\frac{I_{xz}}{I_{xx}}\right)$$

$$C_{\phi R} = -Y_\beta\left(N_{\delta_R}L_r - L_{\delta_R}N_r\right) + Y_{\delta_R}\left(L_r N_\beta - N_r L_\beta\right) + (U_o - Y_r)\left(N_\beta L_{\delta_R} - L_\beta N_{\delta_R}\right)$$

The coefficients for sideslip-angle-to- rudder-deflection transfer function are:

$$A_{\beta R} = Y_{\delta_R}\left(1 - \frac{I_{xz}}{I_{xx}}\frac{I_{xz}}{I_{zz}}\right)$$

$$B_{\beta R} = -Y_{\delta_R}\left(N_r + L_p + N_p\frac{I_{xz}}{I_{xx}} + L_r\frac{I_{xz}}{I_{zz}}\right) + Y_p\left(L_{\delta_R} + N_{\delta_R}\frac{I_{xz}}{I_{xx}}\right) + (Y_r - U_o)\left(L_{\delta_R}\frac{I_{xz}}{I_{zz}} + N_{\delta_R}\right)$$

$$C_{\beta R} = Y_{\delta_R}\left(N_r L_p - N_p L_r\right) + Y_p\left(N_{\delta_R}L_r - L_{\delta_R}N_r\right) + g\cos\theta_o\left(L_{\delta_R} + N_{\delta_R}\frac{I_{xz}}{I_{xx}}\right) + (Y_r - U_o)\left(L_{\delta_R}N_p - N_{\delta_R}L_p\right)$$

$$D_{\beta R} = g\cos\theta_o\left(N_{\delta_R}L_r - L_{\delta_R}N_r\right)$$

The coefficients for yaw-angle-to- rudder-deflection transfer function are:

$$A_{\psi R} = U_o\left(N_{\delta_R} + L_{\delta_R}\frac{I_{xz}}{I_{ZZ}}\right)$$

$$B_{\psi R} = U_o\left(L_{\delta_R}N_p - N_{\delta_R}L_p\right) - Y_\beta\left(N_{\delta_R} + L_{\delta_R}\frac{I_{xz}}{I_{zz}}\right) + Y_{\delta_R}\left(L_\beta\frac{I_{xz}}{I_{zz}} + N_\beta\right)$$

$$C_{\psi R} = -Y_\beta\left(L_{\delta_R}N_p - N_{\delta_R}L_p\right) + Y_p\left(N_\beta L_{\delta_R} - L_\beta N_{\delta_R}\right) + Y_{\delta_R}\left(L_\beta N_p - N_\beta L_p\right)$$

$$D_{\psi R} = g\cos\theta_o\left(N_\beta L_{\delta_R} - L_\beta N_{\delta_R}\right)$$

In Sect. 6.4, we mainly explored a pure lateral control, and employed a linear lateral dynamic model. However, in reality, lateral motion is nonlinear and highly coupled with directional motion. In Sect. 6.6, the fundamentals and control process and of a turn is discussed that includes both directional and lateral motions. For accurate analysis of directional and lateral motions, a nonlinear dynamic model and computer simulation are required.

6.5   Turning Flight

In Sect. 6.3, pure directional control; and in Sect. 6.4, pure lateral control were presented. However, directional and lateral motions are highly coupled, and must be analyzed simultaneously. This section is devoted to a flight operation—coordinated turn—that requires control of both lateral and directional motions.

6.5.1   Level Turn Governing Equations

One of the necessary flight operations is to change the aircraft direction/heading which is frequently referred to as a turning flight. There are two techniques for turn: (1) Bank-to-turn, (2) Skid-to-turn. A bank to turn is a frequently employed technique by all modern civil aircraft (e.g., transport and GA aircraft). As the name implies, in a bank to turn operation, the aircraft will bank, in order to change the direction (i.e., turn). This is an efficient method to turn, since it allows the aircraft to follow a circular path with a lowest cost and g-load. The rudder as well as the aileron are employed for a bank-to-turn operation.

However, in a skid-to-turn operation, the aircraft will turn without any bank. This technique is mainly utilized by high-speed missiles, due to their engine requirement to have a sufficient air intake. If these types of air vehicle bank during a turn, there may not be enough air into engine inlet, so the engine may shut off during turn. Due to the high-speed (often supersonic) turning flight, the centrifugal force will move the aircraft out of the circular trajectory; thus, it generates a skid. Only rudder is employed for a skid-to-turn operation.

In this book, only bank to turn (i.e., steady level turn) operation is presented. In a level turn, pilot should employ a combination of lateral, directional, and longitudinal motions. In a steady level turn only the heading angle (ψ) changes while the pitch angle, angle of attack, and the bank angle all remain constant. The steady state value of yaw rate (R_1) is:

$$R_1 = \dot{\psi}_1 \cos \theta_1 \cos \phi_1 \qquad (6.65)$$

where subscript 1 refers to the steady state value, and $\dot{\psi}_1$ represents the rate of turn (perpendicular to the horizontal plane).

In a level turn, the following six forces are present: (1) Lift (L), (2) Weight (W), (3) Thrust (T), (4) Centrifugal force (F_C), (5) Drag (D), (6) Aerodynamic side force (F_y). An aircraft in a level turning flight is depicted in Fig. 6.10. When an aircraft banks to turn, the aircraft has a bank angle (ϕ), so, the lift has two components: (1) The horizontal component (L sin (ϕ)), and (2) Vertical component to balance the aircraft weight.

Since the lift always is perpendicular to the wing, when the aircraft is banking in a turn, the lift is tilted at an angle to the vertical. So, only the vertical component is available to oppose the weight of the aircraft. As a result of this, the angle of attack must

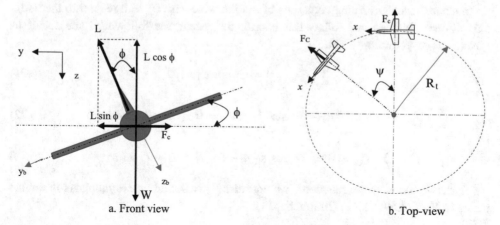

a. Front view   b. Top-view

Fig. 6.10 Level turning flight

be increased to increase the total lift, until its vertical component balances the weight. If this is not achieved, the aircraft will lose altitude during a turn. To increase the angle of attack, up-elevator should be applied.

The force equations to govern a turn are (for vertical and the horizontal plane):

$$\sum F_x = 0 \Rightarrow T \cos \alpha = D \tag{6.66}$$

$$\sum F_z = 0 \Rightarrow L \cos \phi = W \tag{6.67}$$

$$\sum F_y = 0 \Rightarrow L \sin \phi - F_C \pm F_y = 0 \quad \text{(Coordinated turn)} \tag{6.68}$$

$$\sum F_y \neq 0 \Rightarrow L \sin \phi - F_C \pm F_y = ma_y \quad \text{(Un - coordinated turn)} \tag{6.69}$$

The reason for the sign "$\pm$" before side force, F_y (in Eqs. 6.68 and 6.69) is that, depending upon airspeed, this force may need to be positive or negative. In general, in low-speed turns, the side force is positive, while in high-speed turns, the side force is negative. This side force is mainly generated by vertical tail lift, when the rudder is deflected. If there is no lateral acceleration (i.e., $a_y = 0$), the turn is referred to as coordinated. Otherwise, the aircraft will skid outside of the circle, or slip inside.

The centrifugal (outward) force is equal to the aircraft mass multiplied by centripetal (inward) acceleration (a_c):

$$F_C = ma_C = m \frac{U^2}{R_t} \tag{6.70}$$

where R_t denotes the radius of turn.

In Chap. 1, the governing equations of motion were derived with respect to the body-axis system. To keep and follow this coordinate system, the following force equations govern the level turning flight.

$$\sum F_{x_b} = 0 \Rightarrow T \cos \alpha_1 = D \tag{6.71}$$

$$\sum F_{y_b} = 0 \Rightarrow mU_1 R_1 - W \sin \phi_1 \pm F_y = 0 \quad \text{(Coordinated turn)} \tag{6.72}$$

$$\sum F_{z_b} = 0 \Rightarrow W \cos \phi_1 + mU_1 Q_1 = L + T \sin \alpha_1 \tag{6.73}$$

To derive the moment equations, we consider $P_1 = 0$, and ignore moments of engine thrust in Eqs. (1.59)–(1.61) (from Chap. 1):

$$L_{A_1} = (I_{zz} - I_{yy}) R_1 Q_1 \tag{6.74}$$

$$M_1 = I_{xz}(-R_1^2) \tag{6.75}$$

$$N_1 = I_{xz} Q_1 R_1 \tag{6.76}$$

Inserting the n from Eq. (6.13) into to Eq. (6.67), the load factor will be equal to:

$$n = \frac{1}{\cos \phi} \tag{6.77}$$

Hence, the load factor is only a function of the bank angle. This means that as the aircraft bank angle (ϕ) increases, the load factor (n) will increase too. For instance, the load factor in a level turn with a 60° bank angle is equal to two (i.e., cos (60 deg) = 2), which implies that the lift must be twice the aircraft weight.

6.5.2   Coordinated Turn

A coordinated turn is when there is no lateral acceleration (i.e., $a_y = 0$). This turn features: (1) No slipping, (2) No skidding, (3) Constant radius of turn, (4) Constant turn rate, and (5) Fuel is distributed symmetrically. This objective can be achieved with a zero-side force, or a non-zero side force. If the side force kept zero, the centrifugal force will be equal to the horizontal component of the aircraft lift.

In a coordinated turn for a fixed-wing conventional aircraft, all three control surfaces are simultaneously employed: (1) Aileron to roll for creating bank angle, (2) Rudder to yaw for creating a small sideslip angle, (3) Elevator to pitch for increasing angle of attack in order to maintain the altitude.

By expanding aerodynamic forces and moments equations using stability and control derivative, the turn governing equation will make the following form (from simplicity, the subscript 1 is dropped):

Force equations:

$$T \cos \alpha = \frac{1}{2}\rho U^2 S \left(C_{D_o} + C_{D_\alpha}\alpha + C_{D_{\delta_e}}\delta_e\right)$$ (6.78)

$$mUR - W \sin \phi \pm \frac{1}{2}\rho U^2 S \left(C_{y_\beta}\beta + C_{y_r} R\frac{b}{2U} + C_{y_{\delta_a}} + C_{y_{\delta_r}}\delta_r\right) = 0$$ (6.79)

$$W \cos \phi + mUQ = \frac{1}{2}\rho U^2 S \left(C_{L_o} + C_{L_\alpha}\alpha + C_{L_q} Q\frac{C}{2U} + C_{L_{\delta_e}}\delta_e\right) + T \sin \alpha$$ (6.80)

Moment equations:

$$\frac{1}{2}\rho U^2 Sb \left(C_{l_\beta}\beta + C_{l_r} R\frac{b}{2U} + C_{l_{\delta_a}}\delta_a + C_{l_{\delta_r}}\delta_r\right) = (I_{zz} - I_{yy})RQ$$ (6.81)

$$\frac{1}{2}\rho U^2 SC \left(C_{m_o} + C_{m_\alpha}\alpha + C_{m_q} Q\frac{C}{2U} + C_{m_{\delta_e}}\delta_e\right) = I_{xz}(-R^2)$$ (6.82)

$$\frac{1}{2}\rho U^2 Sb \left(C_{n_\beta}\beta + C_{n_r} R\frac{b}{2U} + C_{n_{\delta_a}}\delta_a + C_{n_{\delta_r}}\delta_r\right) = I_{xz}QR$$ (6.83)

Unknowns to these six algebraic equations could be: (1) δ_a, (2) δ_e, (3) δ_r, (4) β, (5) ϕ, (6) T, (7) α, (8) Q, (9) U, and (10) R. Ten unknown variables and only six equations indicate that we need to select three unknowns to determine six unknowns. There are relationships between some of these unknowns that allow us to reduce number of unknowns. The bank angle (ϕ) and airspeed (U) can be selected as to be known.

The steady state pitch rate (Q), and steady state yaw rate (R) are functions of airspeed and bank angle:

$$Q = \frac{g \sin^2(\phi)}{U \cos \phi}$$ (6.84)

$$R = \frac{g \sin(\phi)}{U}$$ (6.85)

The derivation of these two equations is left to the reader as an exercise. Thus, the unknowns of those six equations are two flight angles (α, β), one force (T), and three control surface deflections ($\delta_A, \delta_E, \delta_R$).

In any circular motion, the linear airspeed (here, U) is related to the angular speed (here, $\dot{\psi}$) as:

$$U = R\dot{\psi}$$ (6.86)

A coordinated turn may be achieved with a non-zero side force, or even when the side force is zero. Here, in a coordinated turn, we assume the side force is zero, while there is no lateral acceleration. Thus, from Eq. (6.68):

$$L \sin \phi - F_C = 0 \Rightarrow L \sin \phi = m \frac{U^2}{R_t} = m R_t \dot\psi^2 \tag{6.87}$$

From Eqs. (6.86) and (6.87), and definition of centrifugal force, the *turn radius* (R_t) is determined as a function of airspeed and bank angle:

$$R_t = \frac{U^2}{g \tan \phi} \tag{6.88}$$

The *turn rate* ($\dot\psi$) is found by eliminating the turn radius from Eqs. (4.84) and (4.85):

$$\dot\psi = \frac{g \tan \phi}{U} \tag{6.89}$$

For civil aircraft, standard turn rate is two-minute turn (360 deg), so 3°/s. By considering zero side force for a coordinated turn, we develop three coupled equations (side force equation, a rolling and a yawing moment equation) to solve for a group of unknowns: (1) β, (2) δ_A and (3) δ_R. Then, we replace steady state yaw rate ($R = \frac{g \sin(\phi)}{U}$) with its equivalent from Eq. (6.77) into these three equations:

$$\frac{1}{2}\rho U^2 S \left(C_{y_\beta} \beta + C_{y_r} \frac{g \sin(\phi)}{U} \frac{b}{2U} + C_{y_{\delta_a}} \delta_a + C_{y_{\delta_r}} \delta_r \right) = 0 \tag{6.90}$$

$$\frac{1}{2}\rho U^2 Sb \left(C_{l_\beta} \beta + C_{l_r} \frac{g \sin(\phi)}{U} \frac{b}{2U} + C_{l_{\delta_a}} \delta_a + C_{l_{\delta_r}} \delta_r \right)$$
$$= (I_{zz} - I_{yy}) \frac{g \sin(\phi)}{U} \frac{g \sin^2(\phi)}{U \cos \phi} \tag{6.91}$$

$$\frac{1}{2}\rho U^2 Sb \left(C_{n_\beta} \beta + C_{n_r} \frac{g \sin(\phi)}{U} \frac{b}{2U} + C_{n_{\delta_a}} \delta_a + C_{n_{\delta_r}} \delta_r \right)$$
$$= I_{xz} \frac{g \sin^2(\phi)}{U \cos \phi} \frac{g \sin(\phi)}{U} \tag{6.92}$$

It is interesting to note that these equations are three lateral-directional governing equations of a turning flight. They are decoupled from longitudinal equations and can be reformatted into a matrix form as:

$$\begin{bmatrix} C_{y_\beta} & C_{y_{\delta_a}} & C_{y_{\delta_r}} \\ C_{l_\beta} & C_{l_{\delta_a}} & C_{l_{\delta_r}} \\ C_{n_\beta} & C_{n_{\delta_a}} & C_{n_{\delta_r}} \end{bmatrix} \begin{bmatrix} \beta \\ \delta_a \\ \delta_r \end{bmatrix} = \begin{bmatrix} -C_{y_r} \dfrac{bg \sin(\phi)}{2U^2} \\[2mm] \dfrac{(I_{zz}-I_{yy})g^2 \sin^3(\phi)}{\frac{1}{2}\rho U^4 Sb^2 \cos\phi} - C_{l_r} \dfrac{bg \sin(\phi)}{2U^2} \\[2mm] \dfrac{I_{xz}g^2 \sin^3(\phi)}{\frac{2}{3}\rho U^4 Sb \cos\phi} - C_{n_r} \dfrac{bg \sin(\phi)}{2U^2} \end{bmatrix} \tag{6.93}$$

This is a system of three coupled algebraic equations in matrix form; it is simplified as:

$$[A_{T1}] \begin{bmatrix} \beta \\ \delta_a \\ \delta_r \end{bmatrix} = \begin{bmatrix} C_1 \\ C_2 \\ C_3 \end{bmatrix} \quad (6.94)$$

The Cramer's rule (i.e., ratio of two determinants) may be utilized to solve these equations. The sideslip angle is:

$$\beta = \frac{\begin{vmatrix} C_1 & C_{y\delta_a} & C_{y\delta_r} \\ C_2 & C_{l\delta_a} & C_{l\delta_r} \\ C_3 & C_{n\delta_a} & C_{n\delta_r} \end{vmatrix}}{|A_{T1}|} \quad (6.95)$$

The required aileron deflection for a coordinated turn is:

$$\delta_A = \frac{\begin{vmatrix} C_{y\beta} & C_1 & C_{y\delta_r} \\ C_{l\beta} & C_2 & C_{l\delta_r} \\ C_{n\beta} & C_3 & C_{n\delta_r} \end{vmatrix}}{|A_{T1}|} \quad (6.96)$$

The required rudder deflection for a coordinated turn is:

$$\delta_R = \frac{\begin{vmatrix} C_{y\beta} & C_{y\delta_a} & C_1 \\ C_{l\beta} & C_{l\delta_a} & C_2 \\ C_{n\beta} & C_{n\delta_a} & C_3 \end{vmatrix}}{|A_{T1}|} \quad (6.97)$$

The solution indicates how to adjust the relation between aileron and rudder deflections to follow a pure circular path (i.e., coordinated turn). The control derivative $C_{y\delta_a}$ is normally negligible (i.e., $C_{y\delta_a} = 0$). The side force control derivative $C_{y\delta_r}$ is of major importance in yaw control.

In a slipping turn, either: (1) add more rudder, or (2) Reduce bank angle. In a skidding turn, either: (1) add more bank angle, or (2) Reduce the rudder.

Example 6.5 The twin-turboprop commuter aircraft with a weight of 7000 lb, a wing area of 280 ft^2 is turning with a speed of 200 ft/s at 5000 ft altitude. Other geometry and lateral-directional data of the aircraft are shown below.

$$C_{y\beta} = -0.6 \,\frac{1}{\text{rad}}; \, C_{l\beta} = -0.13 \,\frac{1}{\text{rad}}; \, C_{n\beta} = 0.12 \,\frac{1}{\text{rad}}; \, b = 46\,\text{ft}; \, C_{y_r} = 0.4 \,\frac{1}{\text{rad}}$$

$$C_{y_{\delta a}} = 0; \ C_{l_{\delta a}} = 0.15 \frac{1}{\text{rad}}; \ C_{n_{\delta a}} = -0.0012 \frac{1}{\text{rad}}; \ C_{l_r} = 0.06 \frac{1}{\text{rad}}$$

$$C_{y_{\delta r}} = 0.15 \frac{1}{\text{rad}}; \ C_{l_{\delta r}} = 0.01 \frac{1}{\text{rad}}; \ C_{n_{\delta r}} = -0.07 \frac{1}{\text{rad}}; \ C_{n_r} = -0.2 \frac{1}{\text{rad}}$$

$$I_{yy} = 20{,}000 \ \text{slug.ft}^2; \ I_{zz} = 34{,}000 \ \text{slug.ft}^2; \ I_{xz} = 4400 \ \text{slug.ft}^2$$

a. In order to have a coordinated turn at $+\,45°$ of bank angle, what aileron and rudder deflections are required?

b. What will be the steady-state sideslip angle (in degrees) in this turn?

Solution

At 5000 ft, air density is 0.002048 slug/ft³. The lateral-directional governing equations for a coordinated turn is:

$$\begin{bmatrix} C_{y_\beta} & C_{y_{\delta a}} & C_{y_{\delta r}} \\ C_{l_\beta} & C_{l_{\delta a}} & C_{l_{\delta r}} \\ C_{n_\beta} & C_{n_{\delta a}} & C_{n_{\delta r}} \end{bmatrix} \begin{bmatrix} \beta \\ \delta a \\ \delta r \end{bmatrix} = \begin{bmatrix} -C_{y_r} \frac{bg \ \sin(\phi)}{2U^2} \\ \frac{(I_{zz}-I_{yy})g^2 \ \sin^3(\phi)}{\frac{1}{2}\rho U^4 Sb \ \cos\phi} - C_{l_r} \frac{bg \ \sin(\phi)}{2U^2} \\ \frac{I_{xz}g^2 \ \sin^3(\phi)}{\frac{1}{2}\rho U^4 Sb \ \cos\phi} - C_{n_r} \frac{bg \ \sin(\phi)}{2U^2} \end{bmatrix} \tag{6.93}$$

The coefficients of the left-hand side:

$$[A_{T1}] = \begin{bmatrix} C_{y_\beta} & C_{y_{\delta a}} & C_{y_{\delta r}} \\ C_{l_\beta} & C_{l_{\delta a}} & C_{l_{\delta r}} \\ C_{n_\beta} & C_{n_{\delta a}} & C_{n_{\delta r}} \end{bmatrix} = \begin{bmatrix} -0.6 & 0 & 0.15 \\ -0.13 & 0.15 & 0.01 \\ 0.12 & -0.0012 & -0.07 \end{bmatrix} \tag{6.94}$$

The coefficients of the right-hand side:

$$\begin{bmatrix} C_1 \\ C_2 \\ C_3 \end{bmatrix} = \begin{bmatrix} -C_{y_r} \frac{bg \ \sin(\phi)}{2U^2} \\ \frac{(I_{zz}-I_{yy})g^2 \ \sin^3(\phi)}{\frac{1}{2}\rho U^4 Sb \ \cos\phi} - C_{l_r} \frac{bg \ \sin(\phi)}{2U^2} \\ \frac{I_{xz}g^2 \ \sin^3(\phi)}{\frac{1}{2}\rho U^4 Sb \ \cos\phi} - C_{n_r} \frac{bg \ \sin(\phi)}{2U^2} \end{bmatrix} = \begin{bmatrix} -0.005 \\ -0.00044 \\ 0.003 \end{bmatrix} \tag{6.95}$$

where

$$C_1 = -C_{y_r} \frac{bg \ \sin(\phi)}{2U^2} = -0.4 \frac{46 \times 32.2 \times \sin(45)}{2 \times (200)^2} = -0.005$$

$$C_2 = \frac{(I_{zz} - I_{yy})g^2 \ \sin^3(\phi)}{\frac{1}{2}\rho U^4 Sb \ \cos\phi} - C_{l_r} \frac{bg \ \sin(\phi)}{2U^2}$$

$$= \frac{(34{,}000 - 20{,}000) \times (32.2)^2 \times \sin^3(45)}{\frac{1}{2} \times 0.002048 \times (200)^4 \times 280 \times 46 \cos(45)} - 0.06 \frac{46 \times 32.2 \sin(45)}{2(200)^2} = -0.00044$$

$$C_3 = \frac{I_{xz}g^2 \sin^3(\phi)}{\frac{1}{2}\rho U^4 Sb \cos\phi} - C_{n_r} \frac{bg \sin(\phi)}{2U^2}$$

$$= \frac{4400 \times (32.2)^2 \times \sin^3(45)}{\frac{1}{2} \times 0.002048 \times (200)^4 \times 280 \times 46 \cos(45)}$$

$$- (-0.2)\frac{46 \times 32.2 \sin(45)}{2(200)^2} = 0.003$$

The sideslip angle:

$$\beta = \frac{\begin{vmatrix} C_1 & C_{y_{\delta a}} & C_{y_{\delta r}} \\ C_2 & C_{l_{\delta a}} & C_{l_{\delta r}} \\ C_3 & C_{n_{\delta a}} & C_{n_{\delta r}} \end{vmatrix}}{|A_{T1}|} = \frac{\begin{vmatrix} -0.005 & 0 & 0.15 \\ -0.00044 & 0.15 & 0.01 \\ 0.003 & -0.001 & -0.07 \end{vmatrix}}{\begin{vmatrix} -0.6 & 0 & 0.15 \\ 0.13 & 0.15 & 0.01 \\ 0.12 & -0.001 & -0.07 \end{vmatrix}}$$

$$= \frac{-6.33 \times 10^{-6}}{0.004} = -0.002 \text{ rad} = -0.1 \text{ deg} \tag{6.95}$$

Aileron deflection:

$$\delta_A = \frac{\begin{vmatrix} C_{y_\beta} & C_1 & C_{y_{\delta r}} \\ C_{l_\beta} & C_2 & C_{l_{\delta r}} \\ C_{n_\beta} & C_3 & C_{n_{\delta r}} \end{vmatrix}}{|A_{T1}|} = \frac{\begin{vmatrix} -0.6 & -0.005 & 0.15 \\ 0.13 & -0.00044 & 0.01 \\ 0.12 & 0.003 & -0.07 \end{vmatrix}}{\begin{vmatrix} -0.6 & 0 & 0.15 \\ 0.13 & 0.15 & 0.01 \\ 0.12 & -0.001 & -0.07 \end{vmatrix}}$$

$$= \frac{-6.03 \times 10^{-6}}{0.004} = -0.002 \text{ rad} = -0.096 \text{ deg} \tag{6.96}$$

Rudder deflection:

$$\delta_R = \frac{\begin{vmatrix} C_{y_\beta} & C_{y_{\delta a}} & C_1 \\ C_{l_\beta} & C_{l_{\delta a}} & C_2 \\ C_{n_\beta} & C_{n_{\delta a}} & C_3 \end{vmatrix}}{|A_{T1}|} = \frac{\begin{vmatrix} -0.6 & 0 & -0.005 \\ 0.13 & 0.15 & -0.00044 \\ 0.12 & -0.001 & 0.003 \end{vmatrix}}{\begin{vmatrix} -0.6 & 0 & 0.15 \\ 0.13 & 0.15 & 0.01 \\ 0.12 & -0.001 & -0.07 \end{vmatrix}}$$

$$= \frac{-1.51 \times 10^{-4}}{0.004} = -0.042 \text{ rad} = -2.4 \text{ deg} \tag{6.97}$$

As the bank angle is increased, the required rudder deflection is increased. For instance, for a $+60°$ bank angle, the rudder should be deflected $-3.1°$ to maintain a coordinated turn.

To calculate the other three unknowns (T, δ_E and α) of a coordinated turn, the other three remaining Eqs. (6.78, 6.80, and 6.82) are employed. These are re-written while R and Q are replaced with their equivalents:

$$T \cos \alpha = \frac{1}{2}\rho U^2 S\left(C_{D_o} + C_{D_\alpha}\alpha + C_{D_{\delta_e}}\delta_e\right) \tag{6.98}$$

$$W \cos \phi + mU \frac{g \sin^2(\phi)}{U \cos \phi}$$
$$= \frac{1}{2}\rho U^2 S\left(C_{L_o} + C_{L_\alpha}\alpha + C_{L_q}\frac{g \sin^2(\phi)}{U \cos \phi}\frac{C}{2U} + C_{L_{\delta_e}}\delta_e\right) + T \sin \alpha \tag{6.99}$$

$$\frac{1}{2}\rho U^2 S C\left(C_{m_o} + C_{m_\alpha}\alpha + C_{m_q}\frac{g \sin^2(\phi)}{U \cos \phi}\frac{C}{2U} + C_{m_{\delta_e}}\delta_e\right) = -I_{xz}\left[\frac{g \sin(\phi)}{U}\right]^2 \tag{6.100}$$

There is no closed-form solution for these nonlinear coupled algebraic equations. In order to find a closed-form solution, we need to make a few assumptions that do not considerably impact the accuracy of the calculations. The following three assumptions are made:

$$T \cos \alpha = T \tag{6.101}$$

$$T \sin \alpha = 0 \tag{6.102}$$

$$C_{L_q}\frac{g \sin^2(\phi)}{U \cos \phi}\frac{C}{2U} = 0 \tag{6.103}$$

By applying these assumptions, we obtain:

$$T = \frac{1}{2}\rho U^2 S\left(C_{D_o} + C_{D_\alpha}\alpha + C_{D_{\delta_e}}\delta_e\right) \tag{6.104}$$

$$W \cos \phi + m\frac{g \sin^2(\phi)}{\cos \phi} = \frac{1}{2}\rho U^2 S\left(C_{L_o} + C_{L_\alpha}\alpha + C_{L_{\delta_e}}\delta_e\right) \tag{6.105}$$

$$\frac{1}{2}\rho U^2 S C\left(C_{m_o} + C_{m_\alpha}\alpha + C_{m_q}\frac{gC \sin^2(\phi)}{2U^2 \cos \phi} + C_{m_{\delta_e}}\delta_e\right) = -I_{xz}\left[\frac{g \sin(\phi)}{U}\right]^2 \tag{6.106}$$

Furthermore, the term $\frac{2mg}{\rho U^2 S}$ is replaced with its equivalent, the lift coefficient, C_L. Then, the engine thrust is modeled as a linear function of engine throttle setting (δ_T) as:

$$T = \delta_T \frac{\partial T}{\partial \delta_T} = \delta_T . C_{\delta_T} \tag{6.107}$$

where C_{δ_T} is a derivative which represents the variations of thrust versus throttle setting. Thus, we obtain:

$$C_{D_\alpha}\alpha + C_{D_{\delta_e}}\delta_e - \frac{2}{\rho U^2 S}\delta_T . C_{\delta_T} = -C_{D_o} \tag{6.108}$$

$$C_{L_\alpha}\alpha + C_{L_{\delta_e}}\delta_e = C_L \cos \phi + C_L \frac{\sin^2(\phi)}{\cos \phi} - C_{L_o} \tag{6.109}$$

$$C_{m_\alpha}\alpha + C_{m_{\delta_e}}\delta_e = -\frac{2I_{xz}g^2 \sin^2(\phi)}{\rho U^4 SC} - C_{m_q}\frac{gC \sin^2(\phi)}{2U^2 \cos \phi} - C_{m_o} \tag{6.110}$$

It is interesting to note that, these equations are three longitudinal governing equations of a coordinated turning flight. They are decoupled from lateral-directional equations and can be reformatted into a matrix form as:

$$\begin{bmatrix} C_{D_\alpha} & C_{D_{\delta_e}} & -\frac{2C_{\delta_T}}{\rho U^2 S} \\ C_{L_\alpha} & C_{L_{\delta_e}} & 0 \\ C_{m_\alpha} & C_{m_{\delta_e}} & 0 \end{bmatrix} \begin{bmatrix} \alpha \\ \delta_E \\ \delta_T \end{bmatrix} = \begin{bmatrix} -C_{D_o} \\ C_L \cos \phi + C_L \frac{\sin^2(\phi)}{\cos \phi} - C_{L_o} \\ -\frac{2I_{xz}g^2 \sin^2(\phi)}{\rho U^4 SC} - C_{m_q}\frac{gC \sin^2(\phi)}{2U^2 \cos \phi} - C_{m_o} \end{bmatrix} \tag{6.111}$$

This is a system of three coupled algebraic equations in matrix form; it is simplified as:

$$[A_{T2}] \begin{bmatrix} \alpha \\ \delta_E \\ \delta_T \end{bmatrix} = \begin{bmatrix} C_4 \\ C_5 \\ C_6 \end{bmatrix} \tag{6.112}$$

The Cramer's rule (i.e., using ratio of two determinants) may be utilized to solve these equations. The angle of attack is:

$$\alpha = \frac{\begin{vmatrix} C_4 & C_{D_{\delta_e}} & -\frac{2C_{\delta_T}}{\rho U^2 S} \\ C_5 & C_{L_{\delta_e}} & 0 \\ C_6 & C_{m_{\delta_e}} & 0 \end{vmatrix}}{|A_{T2}|} \tag{6.113}$$

The necessary elevator deflection for a coordinated turn is:

$$\delta_E = \frac{\begin{vmatrix} C_{D_\alpha} & C_4 & -\frac{2C_{\delta_T}}{\rho U^2 S} \\ C_{L_\alpha} & C_5 & 0 \\ C_{m_\alpha} & C_6 & 0 \end{vmatrix}}{|A_{T2}|} \tag{6.114}$$

The necessary throttle setting to maintain a coordinated turn is:

$$\delta_T = \frac{\begin{vmatrix} C_{D_\alpha} & C_{D_{\delta_e}} & C_4 \\ C_{L_\alpha} & C_{L_{\delta_e}} & C_5 \\ C_{m_\alpha} & C_{m_{\delta_e}} & C_6 \end{vmatrix}}{|A_{T2}|} \tag{6.115}$$

Applying determinant formula, the following equations are obtained:

$$\alpha = \frac{-\frac{2C_{\delta_T}}{\rho U^2 S}\left(C_5 C_{m_{\delta_e}} - C_6.C_{L_{\delta_e}}\right)}{-\frac{2C_{\delta_T}}{\rho U^2 S}\left(C_{L_\alpha} C_{m_{\delta_e}} - C_{m_\alpha}.C_{L_{\delta_e}}\right)} = \frac{C_5 C_{m_{\delta_e}} - C_6.C_{L_{\delta_e}}}{C_{L_\alpha} C_{m_{\delta_e}} - C_{m_\alpha}.C_{L_{\delta_e}}} \tag{6.116}$$

$$\delta_E = \frac{-\frac{2C_{\delta_T}}{\rho U^2 S}\left(C_{L_\alpha} C_6 - C_{m_\alpha}.C_5\right)}{-\frac{2C_{\delta_T}}{\rho U^2 S}\left(C_{L_\alpha} C_{m_{\delta_e}} - C_{m_\alpha}.C_{L_{\delta_e}}\right)} = \frac{C_{L_\alpha} C_6 - C_{m_\alpha}.C_5}{C_{L_\alpha} C_{m_{\delta_e}} - C_{m_\alpha}.C_{L_{\delta_e}}} \tag{6.117}$$

$$\delta_T = \frac{C_{D_\alpha}\left[C_{L_{\delta_e}} C6 - C_{m_{\delta_e}} C5\right] - C_{D_{\delta_e}}\left[C_{L_\alpha} C6 - C_{m_\alpha} C5\right] + C_4\left[C_{L_\alpha} C_{m_{\delta_e}} - C_{m_\alpha} C_{L_{\delta_e}}\right]}{-\frac{2C_{\delta_T}}{\rho U^2 S}\left(C_{L_\alpha} C_{m_{\delta_e}} - C_{m_\alpha}.C_{L_{\delta_e}}\right)} \tag{6.118}$$

The solution indicates how much throttle setting (i.e., thrust) and elevator deflection are needed, and how much is the aircraft angle of attack, to follow a pure circular path.

Example 6.6 Consider the twin-turboprop commuter aircraft introduced in Example 6.5. Other geometry and longitudinal data of the aircraft are shown below.

$$C_{L_o} = 0.3; \; C_{D_o} = 0.08; \; C_{m_o} = 0.1; \; C_{m_q} = -34\,\frac{1}{\text{rad}};$$

$$C_{D_\alpha} = 0.9\,\frac{1}{\text{rad}}; \; C_{L_\alpha} = 6.1\frac{1}{\text{rad}}; \; C_{m_\alpha} = -2\,\frac{1}{\text{rad}}; \; C_{\delta_T} = 5000\,\frac{\text{lbf}}{\text{rad}}$$

$$C_{D_{\delta_e}} = 0; \; C_{L_{\delta_e}} = 0.6\,\frac{1}{\text{rad}}; \; C_{m_{\delta_e}} = -2\,\frac{1}{\text{rad}}$$

a. In order to have a coordinated turn at 45° of bank angle, what elevator deflection is required?
b. What will be the steady-state angle of attack in this turn?
c. How much thrust is required?
d. Determine turn radius, and turn rate, and load factor.

Solution

The longitudinal governing equation for a coordinated turn is:

$$
\begin{bmatrix}
C_{D_\alpha} & C_{D_{\delta_e}} & -\frac{2C_{\delta_T}}{\rho U^2 S} \\
C_{L_\alpha} & C_{L_{\delta_e}} & 0 \\
C_{m_\alpha} & C_{m_{\delta_e}} & 0
\end{bmatrix}
\begin{bmatrix}
\alpha \\
\delta_E \\
\delta_T
\end{bmatrix}
=
\begin{bmatrix}
-C_{D_o} \\
C_L \cos\phi + C_L \frac{\sin^2(\phi)}{\cos\phi} - C_{L_o} \\
-\frac{2I_{xz}g^2 \sin^2(\phi)}{\rho U^4 SC} - C_{m_q}\frac{gC \sin^2(\phi)}{2U^2 \cos\phi} - C_{m_o}
\end{bmatrix}
\tag{6.111}
$$

The lift coefficient:

$$
C_L = \frac{2W}{\rho U^2 S} = \frac{2 \times 11{,}000}{0.002048 \times (200)^4 \times 280} = 0.96
\tag{2.27}
$$

Wing mean aerodynamic chord:

$$
C = \frac{S}{b} = \frac{280}{46} = 6.09 \, \text{ft}
\tag{3.66}
$$

The coefficients of the left-hand side:

$$
[A_{T2}] =
\begin{bmatrix}
C_{D_\alpha} & C_{D_{\delta_e}} & -\frac{2C_{\delta_T}}{\rho U^2 S} \\
C_{L_\alpha} & C_{L_{\delta_e}} & 0 \\
C_{m_\alpha} & C_{m_{\delta_e}} & 0
\end{bmatrix}
=
\begin{bmatrix}
0.9 & 0 & -0.436 \\
6.1 & 0.6 & 0 \\
-2 & -2 & 0
\end{bmatrix}
\tag{6.94}
$$

where

$$
\frac{2C_{\delta_T}}{\rho U^2 S} = \frac{2 \times 5000}{0.002048 \times (200)^4 \times 280} = 0.436
$$

The coefficients of the right-hand side:

$$
\begin{bmatrix}
-C_{D_o} \\
C_L \cos\phi + C_L \frac{\sin^2(\phi)}{\cos\phi} - C_{L_o} \\
-\frac{2I_{xz}g^2 \sin^2(\phi)}{\rho U^4 SC} - C_{m_q}\frac{gC \sin^2(\phi)}{2U^2 \cos\phi} - C_{m_o}
\end{bmatrix}
=
\begin{bmatrix}
-0.08 \\
1.056 \\
-0.042
\end{bmatrix}
$$

where

$$
C_4 = -C_{D_o} = -0.08
$$

$$
C_5 = C_L \cos\phi + C_L \frac{\sin^2(\phi)}{\cos\phi} - C_{L_o} = 0.96\cos(45) + 0.96\frac{\sin^2(45)}{\cos(45)} - 0.3 = 1.056
$$

$$
C_6 = -\frac{2I_{xz}g^2 \sin^2(\phi)}{\rho U^4 SC} - C_{m_q}\frac{gC \sin^2(\phi)}{2U^2 \cos\phi} - C_{m_o} = -\frac{2 \times 4400 \times (32.2)^2 \sin^2(45)}{0.002048 \times (200)^4 \times 280 \times 6.09}
$$

$$
- 34\frac{32.2 \times 6.09 \sin^2(45)}{2 \times (200)^2 \cos(45)} - 0.1 = -0.042
$$

The angle of attack:

$$\alpha = \frac{C_5 C_{m_{\delta e}} - C_6.C_{L_{\delta e}}}{C_{L_\alpha} C_{m_{\delta e}} - C_{m_\alpha}.C_{L_{\delta e}}} = \frac{1.056 \times (-2) - (-0.042) \times 0.6}{6.1 \times (-2) - (-2) \times 0.6} = 0.19 \, \text{rad} = 10.9 \, \text{deg}$$

(6.116)

The elevator deflection:

$$\delta_E = \frac{C_{L_\alpha} C_6 - C_{m_\alpha}.C_5}{C_{L_\alpha} C_{m_{\delta e}} - C_{m_\alpha}.C_{L_{\delta e}}} = \frac{6.1 \times (-0.042) - (-2) \times 1.056}{6.1 \times (-2) - (-2) \times 0.6} = -0.17 \, \text{rad} = -9.7 \, \text{deg}$$

(6.117)

The throttle setting:

$$\delta_T = \frac{\begin{vmatrix} C_{D_\alpha} & C_{D_{\delta e}} & C_4 \\ C_{L_\alpha} & C_{L_{\delta e}} & C_5 \\ C_{m_\alpha} & C_{m_{\delta e}} & C_6 \end{vmatrix}}{\begin{vmatrix} C_{D_\alpha} & C_{D_{\delta e}} & -\frac{2C_{\delta T}}{\rho U^2 S} \\ C_{L_\alpha} & C_{L_{\delta e}} & 0 \\ C_{m_\alpha} & C_{m_{\delta e}} & 0 \end{vmatrix}} = \frac{\begin{vmatrix} 0.9 & 0 & -0.08 \\ 6.1 & 0.6 & 1.056 \\ -2 & -2 & -0.042 \end{vmatrix}}{\begin{vmatrix} 0.9 & 0 & -0.436 \\ 6.1 & 0.6 & 0 \\ -2 & -2 & 0 \end{vmatrix}} = \frac{2.76}{4.79} = 0.57 \, \text{rad} = 32.96 \, \text{deg}$$

(6.118)

The required thrust is:

$$\text{T} = \delta_T.C_{\delta_T} = 0.57 \times 5000 = 2876 \, \text{lbf}$$

(6.107)

turn radius, and turn rate, and load factor

$$R_t = \frac{U^2}{g \tan \phi} = \frac{(200)^2}{32.2 \tan (45)} = 1243 \, \text{ft}$$

(6.88)

$$\dot{\psi} = \frac{g \tan \phi}{U} = \frac{32.2 \tan (45)}{200} = 0.161 \frac{\text{rad}}{\text{s}} = 9.2 \frac{\text{deg}}{\text{s}}$$

(6.89)

$$n = \frac{1}{\cos \phi} = \frac{1}{\cos (45)} = 1.414$$

(6.77)

6.6   Turn Control Phenomena

There are a few phenomena which occurs during a turning flight that need to be controlled. Here, two important phenomena are explored: (1) Adverse yaw, and (2) Aileron reversal.

6.6.1 Adverse Yaw

When an airplane is banked to execute a turn, it is desired that aircraft yaws and rolls simultaneously. Furthermore, it is beneficial to have the yawing and rolling moments in the same direction (i.e., either positive or negative). For instance, when an aircraft is to turn to the right, it should be rolled (about x-axis) clockwise and yawed (about z-axis) clockwise. In such a turn, the pilot will have a happy and comfortable feeling. Such yawing moment is referred to as pro-verse yaw, and such turn is a prerequisite for a coordinated turn. This yaw keeps the aircraft pointing into the relative wind. On the other hand, if the aircraft yaws in a direction opposite to the desired turn direction (i.e., a positive roll, but a negative yaw); pilot will not have a desirable feeling and aircraft turn is not coordinated. This yawing moment is referred to as adverse yaw. The reason the aileron induced yawing moment is called adverse is because, it tends to yaw an aircraft out of an intended turn.

To see why adverse yaw happens, see Fig. 6.11, where pilot is planning to turn to the right. For such a goal, the pilot must apply a positive aileron deflection (i.e., left-aileron down and right-aileron up). The lift distribution over the wing in a cruising flight is symmetric; i.e., the right-wing-section lift and the left-wing-section lift are the same. When the left aileron is deflected down and right aileron is deflected up, the lift distribution varies such that the right-wing-section lift is more than left-wing-section lift. Such deflections create a clockwise rolling moment as desired.

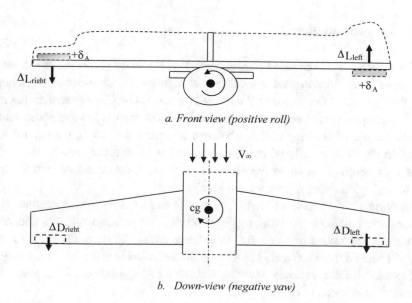

a. Front view (positive roll)

b. Down-view (negative yaw)

Fig. 6.11 Aileron in adverse yaw

However, the aileron deflection simultaneously alters the induced drag of right and left wing differently. Recall that wing drag components of two parts: zero-lift drag (D_o) and induced drag (D_i). The wing induced drag is a function of wing lift coefficient ($C_{D_i} = K \cdot C_L^2$). Since the right-wing-section local lift coefficient is higher than the left-wing-section local lift coefficient, the right-wing-section drag is higher than the left-wing-section drag. The drag is an aerodynamic force and has an arm relative to the aircraft center of gravity. The drag direction is rearward, so this wing-drag-couple is generating a negative (see Fig. 6.11b) yawing moment (i.e., adverse yaw).

The adverse yaw results in an unbalanced condition and needs to be alleviated. Three means for avoiding adverse yaw are: (1) *Spoilers*, (2) *Frise ailerons, and* (3) *Rudder deflection.* Spoilers achieve the desired result by reducing the lift and increasing the drag on the side where the spoiler is raised. Frise ailerons can eliminate adverse yaw by increasing the drag on the side of the upgoing aileron. This is achieved by the choice of hinge location and shaping the aileron nose. When aileron deflected upward, the aileron gap is increased, and a relatively sharp nose protrudes into the stream. These geometrical changes increase the drag.

The adverse yaw can also to be alleviated by the application of rudder deflection. If the rudder is not deflected simultaneously with aileron deflection, the direction of the aileron-generated rolling moment and the wing-drag generated yawing moment would not be coordinated. Thus, when a pilot deflects a conventional aileron to make a turn, the aircraft will initially yaw in a direction opposite to that expected.

6.6.2 Aileron Reversal

Aileron reversal is an undesired phenomenon that is mainly generated due to aeroelastic behavior of wing structure and often occurs at high speeds. Consider the right-section of a wing (Fig. 6.12) with a downward deflected aileron to create a negative rolling moment. At subsonic speeds, the increment lift due to aileron deflection has a centroid somewhere near the middle of the wing chord. However, at supersonic speeds, the control load acts mainly on the deflected aileron itself, and hence has its centroid even farther to the rear. If this load centroid is behind the elastic axis of the wing structure, then a nose-down twist (α_{twist}) of the wing (about y axis) results.

The purpose of this aileron deflection was to raise the right-wing section. However, the wing twist reduces the wing angle of attack, and consequently a reduction of the lift on the right-section of the wing. In extreme cases, the down-lift due to aeroelastic twist will exceed the commanded up-lift, so the net effect is reversed. This change in the lift direction will consequently generate a positive rolling moment. Such phenomenon is referred to as aileron reversal.

This undesired rolling moment implies that the aileron has lost its effectiveness and the roll control derivative; $C_{l_{\delta_A}}$ has changed its sign. This phenomenon poses a significant

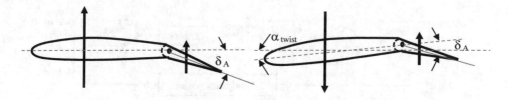

a. An ideal and desired aileron   b. An aileron with aileron reversal

Fig. 6.12 Aileron reversal

constraint on aileron design. In addition, structural design of the wing must examine this aeroelasticity effect of aileron deflection. Most high-performance aircraft have an aileron reversal speed beyond which the ailerons lose their effectiveness, or reverse the outcome.

6.7 Automatic Flight Control Systems

The evolution of modern aircraft in 1950s created a need for automatic-pilot (or autopilot) control systems. The automatic flight control systems (AFCS) are the primary on-board tool for long flight operations and is the foundation for the airspace modernization initiatives. Moreover, technological advances in wireless communication and micro-electro-mechanical systems, make it possible to use inexpensive small autopilots in unmanned aircraft. A significant element in any unmanned aircraft, fighters, and large manned aircraft is the AFCS.

One of the most valuable benefits of using the autopilot is delegating the constant task of manipulating the control surfaces and engine throttle. This benefit allows the pilot more time to manage and observe the entire flight situation.

Autopilot is the integrated software and hardware that serve the three functions of (1) Control, (2) Navigation, and (3) Guidance. In a typical autopilot, three laws are governing simultaneously in three subsystems: (1) Control law in control systems; (2) Guidance law in guidance systems; and (3) Navigation law in navigation systems. In the design of an autopilot, all three laws need to be selected/designed.

In a conventional autopilot, three laws are simultaneously governing the three subsystems: (1) Control system through a control law, (2) Guidance system via a guidance law, and (3) Navigation system through a navigation law. In the design of an autopilot, all three laws need to be developed/designed. The design of the control law is at the heart of autopilot design process. The relation between the control system, the guidance system, and the navigation system is shown in Fig. 6.13. This figure is not illustrating the command system, it will be covered later.

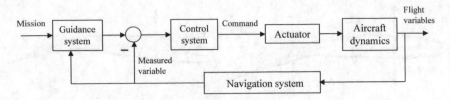

Fig. 6.13 Control, guidance and navigation systems in an autopilot

One of the main subsystems within an AFCS is the control system which is used to keep an aircraft on a predetermined course or heading, necessary for the mission. The control system is using the vehicle state information provided by the on-board sensors to drive the control surface actuators (i.e., servos). In general, a closed-loop control system tends to provide four functions: (1) Regulating, (2) Tracking, (3) Stabilizing, (4) Improve the plant response.

With advances in computer technology, and the introduction of new mathematics theories in nonlinear systems, more applications of advanced control system design techniques such as robust and nonlinear control are seen in the literature. In this Section, a brief review of the fundamentals of closed-loop control system, and flight control requirements are presented.

A graphical illustration of three control functions using three primary control surfaces for a fixed-wing aircraft is sketched in Fig. 6.14. Control is performed mainly by autopilot through moving the control surfaces/engine throttle. The desired change is basically expressed with a reference to the time that takes to move from initial trim point to the final trim point.

There are various measurement devices to measure the flight variables such as airspeed, pitch angle, heading angle, bank angle, linear accelerations, angular rates, altitude, and position. Typical flight measurement devices (sensors) are: (1) Attitude gyroscope, (2) Rate gyroscope, (3) Pitot-tube, (4) Altimeter, (5) Magnetometer, (6) Compass, (7)

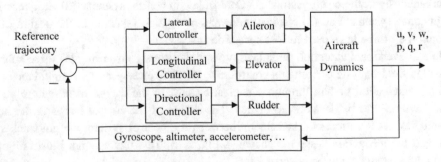

Fig. 6.14 Flight control system with conventional control surfaces

Accelerometer, (8) GPS, (9) Flow incidence angle sensors (angle of attack and sideslip indicators), and (10) Airspeed sensor.

The traditional gyros are made of spinning wheels, but progress in microelectromechanical systems (MEMS) resulted in new effective sensors. MEMS are a new group of low-cost, light-weight sensors for variety of applications from measurement of pressure/temperature to acceleration/attitude. The requirements for a remote pilot in command, or person manipulating the flight controls of a small unmanned aircraft are provided in Part 107 of Federal Aviation Regulations [7]. The author has presented the fundamentals of automatic flight control systems in [8], so the materials are not repeated in this text.

The applications of various flight control systems design techniques and employing different controller such as Proportional-Integral-Derivative (PID), Linear Quadratic Regulator (LQR), dynamic Inversion, model-following, robust output-feedback, and digital controllers have been presented in [9].

6.8 Problems

1. A transport aircraft (An-74) has twin turbofan engines each generating 64 kN of thrust. The distance between each engine thrust line and aircraft center of gravity (along y-axis) is 4.9 m. Other characteristics are:

   $S = 98$ m^2, b = 32 m, $Cn_{\delta r} = -0.12$ 1/rad, $\delta_{Rmax} = \pm 35$ deg.

   Determine the minimum controllable speed at sea level.

2. A transport aircraft has twin turbofan engines each generating 140 kN of thrust. The distance between each engine thrust line and aircraft center of gravity (along y-axis) is 9.6 m, and the distance between vertical tail aerodynamic center to the aircraft center of gravity (l_{vt}) is 38 m. If the maximum lift coefficient of the vertical tail is 1.3 and vertical tail area is 35 m^2, determine minimum control speed of this aircraft at sea level.

3. A light transport aircraft with a mass of 10,000 kg, wing area of 45 m^2 and wing span of 17 m is cruising at 30,000 ft with a speed of 350 knot.

$$C_{l_{\delta_A}} = 0.2\frac{1}{\text{rad}}; C_{l_P} = -0.4\frac{1}{\text{rad}}$$

Estimate the bank angle response of the aircraft to + 5° of aileron.

4. Compare the aircraft angle of attack and elevator deflection for the following aircraft at two different flight conditions:

   a. Cruising flight with a speed of 150 knot—sea level
   b. Level turn (airspeed: 150 knot, Bank angle: 30 deg)—sea level

$$C_{m_\alpha} = -1.5\frac{1}{\text{rad}}; C_{L_\alpha} = 5.5\frac{1}{\text{rad}}; C_{L_{\delta_E}} = 0.6\frac{1}{\text{rad}}; C_{m_{\delta_E}} = -2.2\frac{1}{\text{rad}}; C_{m_q} = -28\frac{1}{\text{rad}}$$

$$W = 5000 \text{ lb}; \ S = 180 \text{ ft}^2; \ b = 38 \text{ ft}; \ I_{xz} = 0; \ C_{L_o} = 0.1 \frac{1}{\text{rad}}; \ C_{m_o} = -0.4$$

5. A single engine GA aircraft with a weight of 2500 lb and a wing area of 150 ft^2 is cruising at sea level with a speed of 84 knot. The aircraft has the following sideslip-angle-to- and yaw-angle-to-rudder-deflection transfer functions for this flight condition:

$$\frac{\beta(s)}{\delta_R(s)} = \frac{32s^3 + 2900s^2 + 32{,}000s - 450}{260s^4 + 3000s^3 + 7100s^2 + 35{,}560s + 480}$$

$$\frac{\psi(s)}{\delta_R(s)} = \frac{-(2600s^3 + 25{,}600s^2 + 2600s + 8100)}{260s^4 + 3000s^3 + 7100s^2 + 35{,}560s + 480}$$

The pilot deflects rudder with the amount of $+12°$. Analyze the yaw response of the aircraft to this step input in rudder deflection.

6. Consider a fixed-wing General Aviation aircraft with a maximum takeoff weight of 2400 lb, a wing planform area of 180 ft^2, is cruising with a speed of 90 knot at 10,000 ft where air density is 17.56×10^{-4} slug/ft^3. Other characteristics are:

$$C_{L_{\alpha_v}} = 5.4 \frac{1}{\text{rad}}; \ b = 39 \text{ ft}; \ \eta_v = 0.95; \ l_v = 16 \text{ ft}; \ S_v = 38 \text{ ft}^2; \ I_{zz} = 2200 \text{ slug.ft}^2$$

The rudder chord is 28% of the vertical tail chord. The pilot applies a $-10°$ of rudder deflection. Determine the initial angular yaw acceleration ($\ddot{\psi}$) as a response of the aircraft to this input.

7. Consider a fighter aircraft with a weight of 35,000 lb and a wing area of 230 ft^2 is flying at sea level with a speed of 140 knot. The aircraft has a wing span of 24 ft, a mass moment of inertia (about x-axis) of 3600 slug.ft^2 and the following non-dimensional lateral stability and control derivatives:

$$C_{l_{\delta_A}} = 0.05 \frac{1}{\text{rad}}; \ C_{l_P} = -0.6 \frac{1}{\text{rad}}$$

Calculate the roll response of the aircraft to a $+20°$ step input in aileron deflection.

8. The twin-turboprop commuter aircraft with a weight of 8000 lb, a wing area of 290 ft^2 is turning with a speed of 190 ft/s at 5000 ft altitude. Other geometry and lateral-directional data of the aircraft are shown below.

$$C_{y_\beta} = -0.5 \frac{1}{\text{rad}}; \ C_{l_\beta} = -0.12 \frac{1}{\text{rad}}; \ C_{n_\beta} = 0.1 \frac{1}{\text{rad}}; \ b = 45 \text{ ft}; \ C_{y_r} = 0.3 \frac{1}{\text{rad}}$$

$$C_{y_{\delta a}} = 0; \ C_{l_{\delta a}} = 0.13 \frac{1}{\text{rad}}; \ C_{n_{\delta a}} = -0.001 \frac{1}{\text{rad}}; \ C_{l_r} = 0.05 \frac{1}{\text{rad}}$$

$$C_{y_{\delta r}} = 0.14 \frac{1}{\text{rad}}; \ C_{l_{\delta r}} = 0.012 \frac{1}{\text{rad}}; \ C_{n_{\delta r}} = -0.06 \frac{1}{\text{rad}}; \ C_{n_r} = -0.17 \frac{1}{\text{rad}}$$

$$I_{yy} = 21{,}000 \text{ slug.ft}^2; \ I_{zz} = 33{,}000 \text{ slug.ft}^2; \ I_{xz} = 4200 \text{ slug.ft}^2$$

a. In order to have a coordinated turn at $+40°$ of bank angle, what aileron and rudder deflections are required?

b. What will be the steady-state sideslip angle (in degrees) in this turn?

9. Consider the twin-turboprop commuter aircraft introduced in Problem 8. Other geometry and longitudinal data of the aircraft are shown below.

$$C_{L_o} = 0.4; C_{D_o} = 0.07 \frac{1}{rad}; C_{m_o} = -0.1; C_{m_q} = -31 \frac{1}{rad};$$

$$C_{D_\alpha} = 0.8 \frac{1}{rad}; C_{L_\alpha} = 5.8 \frac{1}{rad}; C_{m_\alpha} = -1.8 \frac{1}{rad}; C_{\delta_T} = 4000 \frac{lbf}{rad}$$

$$C_{D_{\delta_e}} = 0; C_{L_{\delta_e}} = 0.5 \frac{1}{rad}; C_{m_{\delta_e}} = -2.2 \frac{1}{rad}$$

a. In order to have a coordinated turn at $40°$ of bank angle, what elevator deflection is required?

b. What will be the steady-state angle of attack in this turn?

c. How much thrust is required?

d. Determine turn radius, and turn rate, and load factor.

10. Reference [10] has provided longitudinal-lateral-directional stability and control derivatives (subsonic database) of fighter aircraft F-16 based on wind tunnel test results at NASA Langley and Ames Research center. The following is the lateral-directional state-space model for two inputs (δ_A and δ_R), and three outputs (p, r, and β):

$$\begin{bmatrix} \dot{r} \\ \dot{\beta} \\ \dot{p} \end{bmatrix} = \begin{bmatrix} -0.383 & 4.88 & 0.172 \\ -0.994 & -0.147 & 0.0024 \\ 1.0017 & -13.48 & -1.476 \end{bmatrix} \begin{bmatrix} r \\ \beta \\ p \end{bmatrix} + \begin{bmatrix} 1.487 & -1.53 \\ 0.0074 & 0.021 \\ -12.01 & 2.1096 \end{bmatrix} \begin{bmatrix} \delta_A \\ \delta_R \end{bmatrix}$$

Simulate and analyze the roll and yaw response of the aircraft, when a $+10°$ of aileron and a $-5°$ of rudder deflections are applied simultaneously (for 2 s).

11. Consider a fighter aircraft with a weight of 2000 lb and a wing area of 160 ft^2 is flying at sea level with a speed of 100 knot. The aircraft has the following dimensional directional stability and control derivatives:

$$N_{\delta_R} = -5.4 \frac{1}{s^2}; N_r = -0.6 \frac{1}{s}; N_\beta = 3.5 \frac{1}{s^2}$$

Simulate the yaw angle response of the aircraft to a $+10°$ step input in rudder deflection.

12. A large transport aircraft has twin turbofan engines each generating 70,000 lb of thrust. The distance between each engine thrust line and aircraft center of gravity (along y-axis) is 38 ft. Other characteristics are:

$S = 3900$ ft^2, b $= 198$ ft, $C-n_{\delta_r} = -0.15$ 1/rad, $\delta_{Rmax} = \pm 30$ deg

Determine the minimum controllable speed at sea level.

References

1. Jackson P., Jane's All the World's Aircraft, Jane's information group, Various years
2. Moorhouse, D., MIL-F-8785C, Military Specification: Flying Qualities of Piloted Airplanes, US Department of Defense, OH, 1980
3. Hoak D. E., Ellison D. E., et al, USAF Stability and Control DATCOM, Flight Control Division, Air Force Flight Dynamics Laboratory, Wright-Patterson AFB, Ohio, 1978
4. Nelson R., Flight Stability and Automatic Control, McGraw Hill, 1989
5. MIL-STD-1797, Flying Qualities of Piloted Aircraft, Department of Defense, Washington DC, 1997
6. Stengel, R., Flight Dynamics, Princeton University, 2004
7. Federal Aviation Regulations, Part 107, Operation and Certification of Small Unmanned Aircraft Systems, Federal Aviation Administration, Department of Transportation, Washington DC, 2016
8. Sadraey M., Automatic Flight Control Systems, Morgan & Claypool, 2020
9. Stevens B. L., Lewis F. L., E. N. Johnson, Aircraft control and simulation, Third edition, John Wiley, 2016
10. Snell, A., F. Enns, and W. Garrard Jr, Nonlinear Inversion Flight Control for a Supermaneuverable Aircraft, *Journal of Guidance, Control and Dynamics,* Vol 15, No. 4, 1992.
11. Federal Aviation Regulations, Part 23, Airworthiness Standards: Normal, Utility, Aerobatic, and Commuter Category Airplanes, Federal Aviation Administration, Department of Transportation, Washington DC
12. Federal Aviation Regulations, Part 25, Airworthiness Standards: Transport Category Airplanes, Federal Aviation Administration, Department of Transportation, Washington DC

Printed in the United States
by Baker & Taylor Publisher Services